Bread Made from Yuca

Selected Chronicles of

Indo-Antillean Cultivation and Use of Cassava

1526–2002

PLATE 5
08.16.2001, BARRANQUITAS
From CAZABI: *Gift of the Americas*, Volume IV of *12/12*

BREAD MADE FROM YUCA

Selected Chronicles of

Indo-Antillean Cultivation and Use of Cassava

1526–2002

with Spanish and English translations

Edited by

Jane Gregory Rubin and Ariana Donalds

INTERAMERICAS®
City of New York 2003

Cover: *Cassava leaf.*

06.15.2002, Isabela, *Boriquén* (Puerto Rico)

© InterAmericas® 2003
Society of Arts and Letters of the Americas
Sociedad de Artes y Letras de las Américas
Société des Arts et Lettres des Amériques
162 East 78th Street, New York, NY, USA 10021

Bread Made from Yuca: *Selected Chronicles of Indo-Antillean Cultivation and Use of Cassava 1526–2002* is a limited special edition of 1500, and accompanies "Marisol Villanueva: The New Old World/*El nuevo viejo mundo*," an exhibition of photographs by Marisol Villanueva, selected from the archives of Phase I of The New Old World/*El nuevo viejo mundo*, at CCA7, Centre for the Contemporary Arts in Port of Spain, Trinidad, that includes materials from an exhibition, "The New Old World: Antilles – Living Beyond the Myth," originally mounted in 2002 at the George Gustav Heye Center of the Smithsonian National Museum of the American Indian, and housed online in the NMAI Conexus™ at www.conexus.si.edu/now; and from Cazabi: *Gift of the Americas*, Volume IV of *12/12*, co-published by InterAmericas and the Research Institute for the Study of Man (RISM) in 2002. This book contains reprints from previously published materials.

Spanish translations by María José Iino, Trish O'Kane, Marisol Villanueva
English translations by Aliza Ali, Earl Raymond Hewitt, Theodore Terrones

Photographs (1999–2002) by Marisol Villanueva
Photograph page 40: Courtesy James Pepper Henry

General Editor: Jane Gregory Rubin
Managing Editor: Ariana Donalds
Editors: Aliza Ali, Carmen-Alicia Fernández, María José Iino, Megan McFarland, Trish O'Kane
Designed by Abigail Sturges

ISBN: 1-892321-03-3
Library of Congress Control Number: 2003111647

Printed by Capitol Offset Company Inc. on Lustro Enamel Dull
Set in Goudy

Printed in the United States of America

Cassava is: food that sustains life; sweet and sour liquids that serve as honey and vinegar; a stew that is eaten and enjoyed by the Indians; firewood from the branches when there is no other; and a potent and deadly poison. . . . Another feature of the cassava bread that comes to mind, one that I was not aware of when this first part of these histories was printed, is that in a certain region of the mainland, excellent wine is made from cassava bread. . . . Therefore, there are seven noteworthy characteristics of cassava.

—Gonzalo Fernández de Oviedo y Valdés

[La yuca es:] alimento para sustentar la vida, licores dulce y agrio que sirven como miel y vinagre, potaje que comen y gustan los indios, ramas para leña cuando no hubiere otras, y veneno tan potente y malo. . . . Otra particularidad que se me ocurre del cazabe que yo no conocía la primera vez que se imprimió esta primera parte de estas historias es que, a su vez, en alguna parte del continente se hace muy buen vino del mismo. . . . Así que son siete las características notables que se encuentran en la yuca.

—Gonzalo Fernández de Oviedo y Valdés

Contents/Índice

CAZABI: *Gift of the Americas*, Volume IV, *12/12*

Plate Captions

The 2003 exhibition "MARISOL VILLANUEVA: THE NEW OLD WORLD/*El nuevo viejo mundo*" held at *Espacio/Espace* InterAmericas Space at CCA7, Centre for the Contemporary Arts, from August 15 to October 17, 2003, contains images from CAZABI: *Gift of the Americas*, Volume IV of *12/12*, a commemorative series of portfolios of *giclée* (Iris) prints. Volume IV was co-published by InterAmericas and the Research Institute for the Study of Man (RISM) in 2002.

Following are the captions from the twelve plates, which use excerpts from *Natural History of the West Indies* (1526), a preliminary summary of *Historia general y natural de las Indias, Islas y Tierra firme del mar Océano* (1535–1557), the master work by sixteenth-century Spanish chronicler Gonzalo Fernández de Oviedo y Valdés (1478–1557). The 1526 summary was translated and edited by Sterling A. Stoudemire in 1959, and appeared as Number 32 of the University of North Carolina's *Studies in the Romance Languages and Literature*. The Fernández de Oviedo text appears below in ***bold italics***. An asterisk denotes a plate that appears in this publication, BREAD MADE FROM YUCA: *Selected Chronicles of Indo-Antillean Cultivation and Use of Cassava 1526–2002*. Two asterisks denote a plate that appears in MARISOL VILLANUEVA: THE NEW OLD WORLD/*El nuevo viejo mundo*, which is reprinted with minimal changes on pages 39–67, with an additional photograph on page 67. (See Appendix C, page 121, for a list of photographic credits and page references.)

PLATE 1
02.21.2000, CACIQUE
Rosa Maria Gómez weaving an *haigana* (basket for cassava).
Rosa Maria Gómez tejiendo una *haigana* (canasta para meter la yuca).

PLATE 2
06.15.2002, ISABELA
Nestor A. Ramos Valle cutting the seed stalk.
Nestor A. Ramos Valle cortando el tallo para la siembra.

"To propagate this plant, the Indians break a branch of it into pieces about two spans long.

PLATE 3*
06.15.2002, ISABELA
Placing and pressing the cassava stalks.
Colocando y presionando los tallos de la yuca.

Other Indians . . . simply level the soil and insert these cuttings at regular intervals in the earth.

PLATE 4*
06.16.2002, ISABELA
Bitter cassava plant.
Planta de yuca amarga.

The leaf is almost like that of hemp, like the palm of a man's open hand with the fingers extended.

PLATE 5*
08.16.2001, BARRANQUITAS
After ten months, the cassava is suitable for harvesting.
Después de diez meses la yuca está lista para la cosecha.

The fruit, which grows among the roots, is in the form of ears resembling large carrots. . . .

PLATE 6
08.27.1999, CACIQUE
Soon after the cassava is harvested, the roots are peeled.
Poco después de la cosecha de la yuca, se pelan las raíces.

Inside, the fruit is white.

PLATE 7*
02.01.2000, ARIMA
Christiana Augustus grating cassava.
Christiana Augustus rallando la yuca.

In order to make bread of it, which is called [cazabi], the Indians grate it

PLATE 8
02.01.2000, ARIMA
Ricardo Bharath-Hernandez placing the pulp inside the *coulevre* (press).
Ricardo Bharath-Hernandez colocando la pulpa dentro del *coulevre* (prensa).

and then press it in a strainer, which is a sort of sack about ten palms or more in length and as big as a man's leg. The Indians make this bag from palms which are woven together as if they were rushes.

PLATE 9*
06.16.2002, ISABELA
Separating the starch from the toxic liquid.
Separando el almidón del líquido tóxico.

By twisting the strainer . . . the juice is extracted from the yuca. This juice is a powerful and deadly poison. . . .

PLATE 10**
02.01.2000, ARIMA
Julie Calderon separating the *catebi* (dried lumps of cassava) in the *manári* (sifter).
Julie Calderon separando el *catebi* (masa de yuca seca) en el *manári* (cedazo).

The residue after the liquid is removed, which is something like moist bran,

PLATE 11*
02.01.2000, ARIMA
Cooking cassava bread.
Cocinando el pan de casabe.

is cooked in the fire on a very hot flat clay vessel of the size they want the loaf to be.

PLATE 12
02.18.2000, CACIQUE
Sanny Yohani Durán works packing cassava bread after school.
Al terminar de su día escolar, Sanny Yohani Durán trabaja empacando el pan de casabe.

When the loaf has become firm it is removed from the fire and cured. This cassava bread can be kept for a year or more. . . ."

Note to the Text

On the issues of contemporary English-language usage and translations to or from Spanish of terms relating to autochthonous Caribbean societies and their physical and cultural descendants.

The editors have awaited, with great interest, *Autochthonous Societies*, volume I of the UNESCO-sponsored *General History of the Caribbean*, published in the Spring of 2003, in the hope that the volume would resolve, to some extent, English-language characterizations relating to such societies and their physical and cultural descendants. The volume, in part translated from Spanish, indicates that designations are still in flux, and variants exist between English usages in Canada, the United States of America, the British West Indies, and the United Kingdom. Accordingly, the editors of this publication have adopted the following usages for this book for the insular Caribbean, where there is also significant literature on the subject in Spanish, French, Dutch, and Portuguese:

The term "autochthonous" is generally used in the English texts to describe the peoples mentioned in European chronicles from the fifteenth and sixteenth centuries, at the point of first contact. If speaking of persons who identify with autochthonous cultures, post influences from Europe, Africa, and Asia, the term used is "Amerindian." The Early Spanish "*los indios*" has been translated as "Indians." To follow present-day Latin American Spanish usage, "*indígena*" has been translated as "indigenous," "*indígenes*" has been translated as "indigenous people," "*los indios*" as Amerindian, and "*raza india*" as "Amerindian heritage."

"Carib" is used only if the contemporary indigenous community describes itself as such. "Taíno" is used variously to refer to the autochthonous people of the Greater Antilles c.1500, to a language, and to indigenous and/or culturally distinct Amerindian communities of the Greater Antilles, including the Dominican Republic, Cuba, the Commonwealth of Puerto Rico, and members of these communities living elsewhere.

In terms of geography, "Indo-Antillean" refers to the insular Caribbean and the northern part of South America, including the Orinoco River Valley. In some cases, Amerindian place-names c.1500, with modern place-names in parentheses, have been used for certain islands. They are *Boriquén* (Puerto Rico), *Cairi* (Trinidad), *Cuba* (Cuba), *Quisqueya* (Hispaniola), and *Waitukubuli* (Dominica).

Introduction

Every project undertaken by InterAmericas/Society of Arts and Letters of the Americas/*Sociedad de Artes y Letras de las Américas/Société des Arts et Lettres des Amériques* seeks to broaden the collective awareness of the Americas. An extraordinary range of flora and fauna evolved over millions of years in the continents of North and South America and the insular Caribbean, most with a very recent history of understanding and implementation by man. This publication, *BREAD MADE FROM YUCA: Selected Chronicles of Indo-Antillean Cultivation and Use of Cassava 1526–2002*, incorporates contemporary materials with a narrative written by Spanish historian Gonzalo Fernández de Oviedo y Valdés (1478–1557) almost 500 years ago, when the Europeans encountered and began to document the autochthonous peoples of the insular Caribbean, who had developed profound knowledge of and respect for their natural environment.

Fernández de Oviedo's work is the most celebrated example of the special attention paid by the early European explorers to unknown plant life, particularly foodstuffs, which would enable further exploration and provision return voyages. Chief among these early botanical "discoveries" was *Manihot esculenta* or "manioc," generally known in the English-speaking Caribbean as "cassava" and in the Spanish-speaking Caribbean as "yuca." The most recent work in this publication includes selected photographs and text pertaining to cassava from the archives of Phase I (The Caribbean: 1999–2002) of Puerto Rican artist Marisol Villanueva's long-term curatorial project, THE NEW OLD WORLD/*El nuevo viejo mundo*, previously published in an exhibition catalog, titled *MARISOL VILLANUEVA: THE NEW OLD WORLD/El nuevo viejo mundo*. These selected "Chronicles" are meant to evoke a sense of time and individual points of view in regard to the subject matter.

Long-term documentation projects that incorporate information-gathering techniques from the social sciences in the creation of art in all media are of special interest to InterAmericas. InterAmericas was first introduced to Villanueva's work documenting the people, traditions, and landscapes of the indigenous peoples in the Spanish or formerly Spanish Americas in 1999, when THE NEW OLD WORLD/*El nuevo viejo mundo* came under the sponsorship of the New York Foundation for the Arts.

To date, selections from the archival materials of Phase I of THE NEW OLD WORLD/*El nuevo viejo mundo* (which document sites in Cuba, Dominica, the Dominican Republic, Puerto Rico, and Trinidad) have been the subject of two exhibitions in New York City: "THE NEW OLD WORLD: Antilles – Living Beyond the Myth" at the Smithsonian National Museum of the American Indian's George Gustav Heye Center (NMAI) from November 8, 2002 to April 20, 2003, and housed online in the NMAI Conexus™ at www.conexus.si.edu/now; and "CAZABI: Gift of the Americas," created and installed by InterAmericas in the Second Floor Gallery of the Research Institute for the Study of Man (RISM) in March 2003. The RISM exhibit features the twelve images of CAZABI: *Gift of the Americas*, Volume IV of *12/12*, a twelve-volume commemorative series of portfolios of twelve *giclée* (Iris) prints published by InterAmericas and various co-publishers. The subject of each volume of *12/12* relates to interests and program activities of InterAmericas since its founding in 1992 as a project of the New York Foundation for the Arts. (The exhibition at RISM also incorporates traditional woven objects used in the processing of cassava and preparation of cassava bread, fabricated by Ricardo Bharath-Hernandez, Chief of the Santa Rosa Carib Community in Arima, Trinidad.)

BREAD MADE FROM YUCA is being published to complement an exhibition, "MARISOL VILLANUEVA: THE NEW OLD WORLD/*El nuevo viejo mundo*," which combines the above-mentioned exhibitions. "MARISOL VILLANUEVA: THE NEW OLD WORLD/*El nuevo viejo mundo*" is being presented in the *Espacio/Espace* InterAmericas Space in CCA7 (an international art center run by Caribbean Contemporary Arts in Port of Spain, Trinidad) along with an exhibition in the Main Gallery of CCA7, "PLACE OF BEGINNINGS: The World Views of The Amerindians of *Cairi* and of Medieval Europe," by John Stollmeyer. Also on view is the monumental painting by Tina Spiro, *The Seville Diptych* (1992), which *inter alia*, depicts Jamaica at the moment of the Spaniards' arrival. These and an exhibition from the collections of the Santa Rosa Carib Community and The Archaeology Centre, Department of History at the University of the West Indies/St. Augustine are on view from August 15 to October 17, 2003.

The twelve *giclée* prints by Marisol Villanueva in CAZABI: *Gift of the Americas* portray the everyday practice of planting and processing cassava as a monumental cultural achievement and spiritual event, where the traditional elements of earth, air, fire, and water are used to create and sustain life, both of the individual and of the community that identifies with the centrality of cassava cultivation.

The black-and-white images and Villanueva texts in BREAD MADE FROM YUCA are from the archives of Phase I and, for the most part, were not included in "THE NEW OLD WORLD: Antilles – Living Beyond the Myth." These photographs, except for the color image of the *pastel* (06.16.2002 ISABELA, page 67), were first published in the exhibition catalog, MARISOL VILLANUEVA: THE NEW OLD WORLD/*El nuevo viejo mundo*. The images are less iconic, more descriptive, and more inclusive than those included in the two previous exhibitions, and were chosen primarily to add context to Villanueva's "chronicles," which are based on her personal journal or *diario*. Except for the image on page 40 by James Pepper Henry and the image of the *pastel*, the color plates in this publication are from CAZABI: *Gift of the Americas*.

The editors have included selections from some of the first partial English translations of the earliest comprehensive European documentation on cassava by Gonzalo Fernández de Oviedo y Valdés: *Historia general y natural de las Indias, Islas y Tierra firme del mar Océano*. Part I, Book VII, Chapter 2 and selections from Part II, Book XXIV, Chapter 3 of *Historia general y natural de las Indias* relate to the cultivation and use of cassava in Hispaniola (today shared by the Dominican Republic and the Republic of Haiti), an island documented in Phase I of the archives of THE NEW OLD WORLD/*El nuevo viejo mundo*; and the Huyapari (Orinoco) River Valley, in northeastern Venezuela, across the Gulf of Paria from Trinidad.

Historia general y natural de las Indias, Islas y Tierra firme del mar Océano was not published in Spanish in its entirety until 1851, and no complete English translation has ever been published. The English translation of the chapter relating to Hispaniola is from an unpublished partial translation done under the auspices of the WPA (Works Progress Administration) in 1941. The English translation of the passages relating to the Orinoco River were commissioned for this publication. The Spanish Fernández de Oviedo texts are extracted from the 1851 publication of *Historia general y natural de las Indias, Islas y Tierra firme del mar Océano*. To document the extraordinary process of preparing what has been described as a "food made from poison," developed in the tropical Americas as early as ten thousand years ago, the twelve plates of CAZABI: *Gift of the Americas* use brief passages from *Natural History of the West Indies*, a 1959 English translation by Sterling A. Stoudemire of Fernández de Oviedo's 1526 preliminary summary of *Historia general y natural de las Indias*. (See pages 9–10 for the plate captions from CAZABI: *Gift of the Americas*; the texts excerpted from the Stoudemire translation appear in ***bold italics***.)

The history of the social and economic importance of the cultivation and use of cassava in the insular Caribbean required context for "*la cultura de la yuca*," or the ritual and social interaction surrounding the plant and its products, including cassava bread, today referred to as *casabe* in Cuba, Puerto Rico, the Dominican Republic, and Venezuela. We are most fortunate to be able to include selected quotations from "*Sobrevivencia de la cultura de la yuca en Puerto Rico*," a chapter from the book *Resistencia y supervivencia indígena en Puerto Rico*, which is a comprehensive study yet unpublished of the continuation of Amerindian cultures in Puerto Rico by scholar Juan Manuel Delgado. A second bilingual publication, containing the complete contents of "*Sobrevivencia de la cultura de la yuca en Puerto Rico*," is in preparation.

Today cassava, or *Manihot esculenta*, is of such international importance that the editors felt it imperative to include "*Manihot Esculenta*: Historic and Economic Context," an essay on cassava's significance written by Trish O'Kane. O'Kane elaborates on recent archaeological studies of the ancient Amerindian use of *Manihot esculenta*, the importance of bread made from yuca to the Early Modern European exploration and settlement of the Americas, and the global adaptation and expansion of cassava production.

In the "Afterword," Stuart W. Lewis, M.D., President and Managing Director of RISM, an institution associated with seminal research on the complex interplay of historic, economic, social, and cultural forces in multiethnic societies, comments on the potential problems associated with the introduction of a new foodstuff into environments that do not share the cultural history associated with its original cultivation and use. This is more than a cautionary tale in the case of cassava, which must be processed carefully to extract its potentially toxic compounds before it can bestow nutritional benefit.

InterAmericas assumes a continuity and influence of the autochthonous cultures of the Antilles on contemporary societies in the insular Caribbean. These societies, through the process of transculturation, have absorbed and redefined influences from Africa, Europe, and Asia (and all points in between) during the past five hundred years.

Botanicals and transculturation are also the subject of Volume V of *12/12* (in publication), a derivative work based on *A Collection of Exotics From the Island of Antigua*, by "A Lady." It is generally thought to date from 1797 by one "Lydia Byam," from a distinguished family in eighteenth-century British America. The twelve images are hand-colored engravings of plants known to be medicinal and cultivated on the island of Antigua at the time. InterAmericas' publications related to *A Collection of Exotics From the Island of Antigua* will include a work on the ethnobotany and the eighteenth-century medicinal use of the plants depicted in the Caribbean, where the healing traditions of the Americas, Africa, and Great Britain were first integrated.

InterAmericas creates projects recognizing the contributions of the many cultures that are in the ongoing process of creating a distinctive modern culture of the Americas. The importance of revisiting the contributions and worldview of the autochthonous cultures as a backdrop for all cultural interventions post 1492 CE cannot be underestimated. The task over the next few years will be not only to review the documentation in text and image by the Europeans arriving in the Americas, but to seek other forms of documentation of the culture of the indigenous peoples of the Americas, derived from the experiences of the Amerindians, and from the Africans and Asians who came or were brought to the "New World" from largely agricultural societies during the fifteenth to nineteenth centuries. This documentation can be found in oral histories, linguistic evidence, and ritualized behavior, particularly involving food and artifacts.

Perhaps one of the reasons that there has been a continuity of *la cultura de la yuca* during the last five hundred years is that the persons arriving in the insular Caribbean from the continents of Africa and Asia seem to have had a shared understanding with the Amerindian communities of a spiritual life centered in nature and of the interrelationship between human beings and their environment.

Jane Gregory Rubin
Director, InterAmericas
City of New York
August 2003

Introducción

Todos los proyectos emprendidos por InterAmericas buscan ampliar la conciencia colectiva de las Américas. Una variedad extraordinaria de flora y fauna evolucionó durante millones de años en América del Norte, América del Sur y el Caribe insular, la mayoría con una historia muy reciente de comprensión y aprovechamiento por parte del hombre. La presente publicación, *BREAD MADE FROM YUCA: Selected Chronicles of Indo-Antillean Cultivation and Use of Cassava 1526–2002*, combina materiales contemporáneos con un relato del historiador español Gonzalo Fernández de Oviedo y Valdés (1478–1557), escrito hace casi 500 años, cuando se produjera el encuentro de los europeos con los pueblos autóctonos del Caribe insular. Entonces se inició la documentación sobre estos últimos, quienes habían desarrollado un conocimiento profundo y un respeto hacia su medio natural.

El trabajo de Fernández de Oviedo es el ejemplo más célebre de la atención especial que los primeros exploradores europeos prestaron a las plantas, en particular a aquellas que eran alimenticias, que les permitirían explorar aún más el territorio y aprovisionarse para los viajes de regreso. El más importante de estos "descubrimientos" botánicos del comienzo fue el de la *Manihot esculenta* o "mandioca", conocida comúnmente en el Caribe anglófono como "cassava" y en el Caribe hispanoparlante como "yuca". El trabajo más reciente en esta publicación incluye un texto y fotografías relacionados con la yuca, que forman parte de los archivos de la Fase I (El Caribe: 1999–2002) del proyecto curatorial a largo plazo de la artista puertorriqueña Marisol Villanueva, THE NEW OLD WORLD/*El nuevo viejo mundo*, publicado en *MARISOL VILLANUEVA: THE NEW OLD WORLD/El nuevo viejo mundo*, un catálogo de exposición. Esta selección de "crónicas" tiene como objeto evocar un sentido del tiempo y puntos de vista individuales en relación al tema que nos ocupa.

Para InterAmericas resultan de especial interés aquellos proyectos de documentación a largo plazo, que incorporan técnicas de recopilación de información de las ciencias sociales en la creación de arte en todos los medios. El primer contacto de InterAmericas con el trabajo de Villanueva surgió, en 1999, a través de la documentación de la población, las tradiciones y los paisajes de los pueblos indígenas en la América Hispana o ex Hispana, cuando THE NEW OLD WORLD/*El nuevo viejo mundo* obtuvo el auspicio de la New York Foundation for the Arts.

Hasta el presente, los materiales escogidos del archivo de la Fase I de THE NEW OLD WORLD/*El nuevo viejo mundo* (que documenta sitios en Cuba, Dominica, la República Dominicana, Puerto Rico, y Trinidad) han sido parte de dos exposiciones en la ciudad de Nueva York: "THE NEW OLD WORLD: Antilles – Living Beyond the Myth" en el George Gustav Heye

Center del Smithsonian National Museum of the American Indian (NMAI), del 8 de noviembre de 2002 al 20 de abril de 2003, y *online* en el sitio del NMAI Conexus™ (www.conexus.si.edu/now); y "CAZABI: Gift of the Americas", creado e instalado por InterAmericas en la Galería del Segundo Piso del Research Institute for the Study of Man (RISM) en marzo de 2003. La exposición del RISM exhibe las doce imágenes de CAZABI: *Gift of the Americas*, Tomo IV de *12/12*, una serie conmemorativa de doce tomos de portafolios que contienen doce impresiones *giclée* (Iris), publicadas por InterAmericas y varios co-editores. El tema de cada tomo de *12/12* está ligado a los intereses y los programas de actividades de InterAmericas desde que ésta se fundara en 1992, como un proyecto de la New York Foundation for the Arts. (La exposición del RISM también incorpora objetos tejidos tradicionales utilizados en el procesamiento de la yuca y en la elaboración del pan de yuca, fabricados por Ricardo Bharath-Hernandez, el jefe del Santa Rosa Carib Community en Arima, Trinidad).

La publicación de *BREAD MADE FROM YUCA* complementa la exposición "MARISOL VILLANUEVA: THE NEW OLD WORLD/*El nuevo viejo mundo*", en la que se combinan selecciones de las exposiciones mencionadas anteriormente. "MARISOL VILLANUEVA: THE NEW OLD WORLD/*El nuevo viejo mundo*" se presenta en el *Espacio/Espace* InterAmericas en CCA7 (un centro de arte internacional a cargo de Caribbean Contemporary Arts en Puerto España, Trinidad) conjuntamente con una exposición en la Galería Principal de CCA7, titulada "PLACE OF BEGINNINGS: The World Views of The Amerindians of *Cairi* and of Medieval Europe" por John Stollmeyer. Asimismo, se ha incluido el cuadro monumental de Tina Spiro titulado *The Seville Diptych* (1992) que, en otras cosas, describe a Jamaica al momento de la llegada de los españoles. Estos trabajos y materiales del Santa Rosa Carib Community y otras podrán ser vistos del 15 de agosto al 17 de octubre de 2003.

Las doce impresiones *giclée* de Marisol Villanueva en CAZABI: *Gift of the Americas* retratan la práctica diaria de plantar y procesar la yuca como logro cultural y acto espiritual monumentales, donde los elementos tradicionales de tierra, aire, fuego, y agua se utilizan para crear y sostener la vida, tanto del individuo como de la comunidad que se identifica con la centralidad del cultivo de la yuca.

Las imágenes en blanco y negro y los textos de Villanueva que se encuentran en *BREAD MADE FROM YUCA* pertenecen a los archivos de la Fase I y en su mayor parte no fueron incluidos en la exposición "THE NEW OLD WORLD: Antilles – Living Beyond the Myth". Estas imágenes, con excepción de la imagen en color del pastel (06.16.2002 ISABELA, página 67), fueron publicadas por primera vez en el catálogo *MARISOL VILLANUEVA: THE NEW OLD WORLD/El nuevo viejo mundo*. Las imágenes son menos icónicas, más descriptivas, y más completas que las incluidas en las dos exposiciones anteriores y fueron seleccionadas básicamente para proporcionar un contexto a la "crónica" de Villanueva, basada en su diario personal. Con excepción de la fotografía en página 40 de James Pepper Henry, y la del pastel, las placas en color de este catálogo pertenecen a CAZABI: *Gift of the Americas*.

Las editores han incluido segmentos de algunas de las primeras traducciones parciales al inglés de la primera documentación detallada europea, escrita por Gonzalo Fernández de Oviedo y Valdés: *Historia general y natural de las Indias, Islas y Tierra firme del mar Océano*. Los textos incluídos de la Primera parte, Libro VII, Capítulo 2 y selecciones de la Segunda parte, Libro XXIV, Capítulo 3 de la *Historia general y natural de las Indias* se refieren al cultivo y uso de la yuca en Hispaniola (isla compartida hoy por la República Dominicana y la República de Haití), documentada en la Fase I de los archivos de THE NEW OLD WORLD/*El nuevo viejo mundo*, y en el Valle del Río Huyapari (Orinoco) al noreste de Venezuela.

Historia general y natural de las Indias, Islas y Tierra firme del mar Océano no fue publicado en forma completa en español sino hasta 1851. Nunca se había publicado una traducción completa al inglés. Las traducciones al inglés de los pasajes relativos a la Hispaniola pertenecen a una traducción parcial no publicada, realizada en 1941 bajo los auspicios de la WPA (siglas en inglés de Works Progress Administration). La traducción de los pasajes relativos al Río Orinoco fue comisionada por los editores de este libro. Los textos de Fernández de Oviedo en español son extractos de una edición en 1851 de la *Historia general* y de elaboración de lo que se ha descrito como un "alimento hecho de veneno", desarrollado en el Trópico de las Américas desde hace al menos 10.000 años, las doce placas de CAZABI: *Gift of the Americas* incluyen breves pasajes de *Natural History of the West Indies* una traducción al inglés, realizada en 1959, por Sterling A. Stoudemire, del resumen preliminar de la obra de Fernández de Oviedo escrito en 1526. (Ver páginas 9–10 para los títulos de las placas de CAZABI: *Gift of the Americas*).

La historia de la importancia social y económica del cultivo y uso de la yuca en el Caribe insular requirió poner en contexto a "la cultura de la yuca", el ritual e interacción social que rodea a la planta y sus derivados, entre los que se incluye el pan de yuca, conocido hoy como "casabe" en Cuba, Puerto Rico, la República Dominicana y Venezuela. Nos sentimos muy afortunados haber podido incluir una selección de citas de "Sobrevivencia de la cultura de la yuca en Puerto Rico", un capítulo perteneciente a *Resistencia y supervivencia indígena en Puerto Rico*, un ensayo sobre la continuación de las culturas amerindias en Puerto Rico, escrito por el académico Juan Manuel Delgado. Una segunda publicación bilingüe con los contenidos completos de "Sobrevivencia de la cultura de la yuca en Puerto Rico" se encuentra en preparación.

Hoy en día, la importancia internacional de la yuca, o *Manihot esculenta* es tal que los editores consideraron imperativo incluir "*Manihot esculenta*: Su contexto histórico y económico", un ensayo sobre su relevancia, escrito por Trish O'Kane. O'Kane explica los recientes estudios arqueológicos relativos al uso de la *Manihot esculenta* por los antiguos amerindios, la importancia del pan de la yuca en la exploración y el establecimiento en las Américas de los primeros europeos modernos y la adaptación y expansión de la producción de la yuca a nivel mundial.

En el "Epílogo", Stuart W. Lewis, Doctor en Medicina, Presidente y Director Ejecutivo de RISM, una institución asociada a la investigación de los orígenes de la compleja interacción de las fuerzas históricas, económicas, sociales, y culturales en las sociedades multiétnicas, comenta acerca de los problemas potenciales asociados con la introducción de un nuevo alimento en medios donde no se comparte una historia cultural ligada a su cultivo y uso original. En el caso de la yuca, se trata de algo más que un relato aleccionador, ya que la misma ha de procesarse con sumo cuidado para extraer sus componentes potencialmente tóxicos antes de poder beneficiarse con sus características alimenticias.

InterAmericas da por sentada la continuidad e influencia de las culturas autóctonas de las Antillas en las sociedades contemporáneas del Caribe insular. Estas sociedades, a través del proceso de transculturación, han absorbido y redefinido las influencias provenientes de África, Europa, y Asia (y todos los puntos entre medio) durante los últimos quinientos años.

Las plantas y transculturación también son los temas del Tomo V of *12/12*, que está en preparación. Se será de un trabajo derivativo basado en *A Collection of Exotics From the Island of Antigua* firmado por "Una Dama". Se cree que fue escrito en 1797 por un "Lydia Byam", miembro de una familia distinguida de la América inglesa del siglo XVIII. Las doce imágenes son grabados coloreados a mano, de plantas que se saben medicinales, que se cultivaban en la isla de Antigua en ese momento. Las publicaciones de InterAmericas relativas a *A Collection of Exotics From the Island of Antigua* incluirán un trabajo sobre la etnobotánica y el uso medicinal en el siglo XVIII de las plantas representativas del Caribe, donde por primera vez se integraron las tradiciones curativas de las Américas, África y Gran Bretaña.

InterAmericas genera proyectos que reconocen los aportes de las varias culturas involucradas en el proceso continuo de crear una cultura moderna y característica de las Américas. No ha de subestimarse la importancia de retomar los aportes y a la cosmovisión de las culturas autóctonas, como telón de fondo para todas las intervenciones culturales posteriores a 1492 CE. El trabajo para los próximos años será, no sólo revisar la documentación en forma de textos e imágenes recopilada por los europeos llegando en las Américas, sino también buscar otras formas de documentación de la cultura de los pueblos indígenas de las Américas, derivadas de las experiencias de los amerindios y de los africanos y asiáticos, que vinieron o fueron traídos al nuevo mundo entre los siglos XV y XIX, provenientes de sociedades en su mayor parte agrícolas. Este tipo de documentación puede encontrarse en historias orales, evidencia lingüística y hábitos con características de rituales que incluyan, en particular, alimentos y artefactos.

Tal vez, una de las razones por la cual "la cultura de la yuca" haya perdurado a lo largo de los últimos quinientos años sea que aquellos que llegaron al Caribe insular desde los continentes de África y Asia parecen haber compartido con las comunidades amerindias la comprensión tanto de una vida espiritual centrada en la naturaleza, como de la interrelación entre los seres humanos y el medio ambiente.

Jane Gregory Rubin
Directora, InterAmericas
Ciudad de Nueva York
Agosto de 2003

PLATE 11
02.01.2000, ARIMA
From CAZABI: *Gift of the Americas*, Volume IV of *12/12*

Historia general y natural de las Indias, Islas y Tierra firme del mar Océano

Part I, Book VII, Chapter 2

and selections from

Part II, Book XXIV, Chapter 3

Primera parte, Libro VII, Capítulo 2

y selecciones de

Segunda parte, Libro XXIV, Capítulo 3

Gonzalo Fernández de Oviedo y Valdés

Gonzalo Fernández de Oviedo y Valdés (1478–1557), Spanish historian and "first chronicler of the New World," was born in Madrid, Spain, and died in Valladolid, "clutching the keys of the fortress of Santo Domingo," according to Sterling A. Stoudemire. He was attached to the Court of Ferdinand and Isabella at an early age and was a page to the *Infante* Don Juan, the heir apparent to the Spanish throne. Fernández de Oviedo was fourteen when he first met Christopher Columbus. After the *Infante*'s death in 1497, Fernández de Oviedo left for Italy and in 1502 returned, according to Antonello Gerbi, "as a whole hearted disciple of humanism, a lover of fine arts and culture, filled with a reverence for antiquity reborn and the new literature, with his eyes avidly opened to a world expanding prodigiously in time and space...able to emulate Pliny...in becoming himself...the Pliny of the lands across the ocean." In 1513, he was appointed *inter alia* "*Veedor de las fundaciones de oro de la Tierra Firme*" (Inspector of the Goldmines of the Mainland). In 1519, King Charles I appointed Fernández de Oviedo to several positions and commissioned him to write the official history of the New World. The result was *Historia general y natural de las Indias, Islas y Tierra firme del mar Océano*.

Editor's Note

During his "turbulent but distinguished"[1] career, Gonzalo Fernández de Oviedo y Valdés crossed the Atlantic Ocean twelve times. During an interlude in Spain beginning in 1524, he wrote *De la natural hystoria de las Indias* (Natural History of the Indies), more generally referred to as *Sumario de la natural historia de las Indias* in modern editions, published in Toledo in 1526. The *Sumario* was an immediate success, as was *Primera parte de la historia natural y general de las Indias, Islas y Tierra firme del mar Océano,* Part I of his *Historia general y natural de las Indias* (General and Natural History of the Indies), published in Seville in 1535, the year that Fernández de Oviedo was appointed the commandant of the fortress of Santo Domingo. By 1544, translations from the original Castilian into Tuscan, French, German, Latin, Greek, Turkish, and Arabic were in process or under discussion.[2]

Part II of *Historia general y natural de las Indias* was ready for press in 1541, but Fernández de Oviedo was unable to return to Spain until 1546, at which point he did not obtain a license to publish. It is believed that permission was refused in part because of the efforts of Bartolomé de las Casas, as *Historia general y natural de las Indias* conflicted with De las Casas' positions on events in the New World.[3]

In 1556 and 1557, Fernández de Oviedo obtained permission to publish, however, he did not live to see the second part of the manuscript in print. It is also doubtful that King Charles I (later Emperor Charles V), who commissioned the account, ever saw a copy of Part II (*Libro XX. De la segunda parte de la general historia de las Indias*), published in Valladolid in 1557.[4] Various explanations have been offered for the hiatus in publication after Fernández de Oviedo's death, from fire to reluctance to publish an official account that contained criticism of the clergy and conduct of the Spanish in the Indies. In all events, in 1563, the "inquisitor in the City of Seville" was ordered by the King to give "certain books . . . 'written in the hand of Gonzalo Fernández de Oviedo, commandant of the fortress of Santo Domingo on the island of Hispaniola, which deal with the Indies, that have not been seen or examined or authorized for publication' to the Council of the Indies 'with all due diligence and due precaution.'"[5]

Having passed through several owners and repositories, an almost complete version of the text was found in 1775 in the archives of the Ministry of Grace, Justice and the Indies, and publication was contemplated by the government but never effected. Nor was an early nineteenth-century edition by the Academia de la Historia of Madrid realized, in part because of the revolutions in the Spanish Americas. The complete publication was undertaken by La Real Academia de la Historia three centuries after the death of Fernández de Oviedo, under the direction of José Amador de los Ríos. This version, totaling more than twenty-four hundred pages, plus a one-hundred page biography of Fernández de Oviedo by Amador de los Ríos, appeared between 1851 to 1855 as *Historia general y natural de las Indias, Islas y Tierra firme del mar Océano.* It has been described as "the most voluminous work written since the time of Adam."[6]

The publications of 1526, 1535, and 1557 influenced the works of chroniclers and scientists of the sixteenth and later centuries. The extreme rarity of the originals, as well as the cost of the 1851 edition, have been an obstacle to the study of Fernández de Oviedo's *Historia general y natural de las Indias,* as has criticism on a number of grounds by later scholars and literary historians.

A serious hindrance to the study of Fernández de Oviedo is the lack of a complete translation into English of *Historia general y natural de las Indias.* There is currently a great deal of research and publishing on various aspects of Fernández de Oviedo, but to the knowledge of the editors, by the time of publication, there is no complete translation in preparation.

The following pages contain Part I, Book VII, Chapter 2 ("Focusing on the bread known by the Indians as *cazabe*...") from an unpublished and incomplete English translation from the Spanish 1851 edition of *Historia general y natural de las Indias, Islas y Tierra firme del mar Océano,* edited by Amador de los Ríos. This translation was executed by Earl Raymond Hewitt (dates unknown) and Theodore Terrones (dates unknown) under the auspices of the Works Progress Administration in 1941. The manuscript is currently housed in the Bancroft Library of the University of California at Berkeley and is not mentioned in the materials cited by Sterling A. Stoudemire in his 1959 translation of *Sumario de la natural historia de las Indias.*

Trinidadian translator Aliza Ali was commissioned to translate, specifically for this publication, Part II, Book XXIV, Chapter 3 ("Focusing on the Huyapari River located in the Gulf of Paria . . . "), from a modern edition of *Historia general y natural de las Indias,* which was edited by Juan Pérez de Tudela Bueso. The parts of Chapter 3 relating to cassava have been excerpted in the following pages. As part of the process, the Spanish texts have been somewhat revised to reflect present-day Spanish usage. The Hewitt/Terrones English translation has been modified accordingly to correspond with the revised Spanish.

Part I, Book VII, Chapter 2

Focusing on the bread known by the Indians as *cazabe,* which is the second type of foodstuff prepared by the Indians not only on this island of Hispaniola but elsewhere. *Cazabe* is also prepared by *los cristianos* today [1537], some of them using it even more than they do corn, considering it better and making more use of it. This bread is made from a plant called cassava, also known by its Spanish name "yuca."

Let us now touch on another type of foodstuff made by the Indians using cassava on this island of Hispaniola, on the other islands settled by *los cristianos*, and even in some parts of the mainland. The cassava plant consists of several knotty stalks somewhat taller than a man (others much shorter), as thick as approximately two fingers. Its height and thickness depend on whether the soil is fertile or arid and also on the species of the plant. One type of cassava bears a resemblance to the leaf of the hemp plant, similar to a man's hand with the fingers stretched open, however, the cassava leaf is larger and thicker than that of the hemp. Each leaf has seven or nine points or divisions. As I have already mentioned, the stalk is extremely knotty, and the stem is whitish brown in color. One variety of cassava is almost purple with exceptionally green leaves, and it enhances the beauty of the countryside when the land where it is planted is well cared for and free of weeds.

There is yet another type of cassava whose branches and fruit are no different from the one mentioned above, but even though each of its leaves similarly has seven or nine divisions, it is shaped differently. To illustrate this, I have included drawings of each of these. Despite the similarities in the leaves, there are, however, varying types of cassava, some greener than others, some with thicker branches than others, the whiteness of the stem may vary, plus, there are differences in terms of the skin, which are not really relevant in this case.

To set this plant (any of the varieties mentioned), a few round mounds of earth are formed in rows, just as grapevines are planted in the kingdom of Toledo, and especially in Madrid, where the vines are set out in rows. Each mound is eight or nine feet in circumference, and the edges of one come close to the next, with little space between them. The top of the mound is not pointed but almost flat, and the highest part extends about knee-high or a little higher. Six, eight, ten, or more pieces of the stem or branch of the cassava plant itself are placed in each mound, approximately two inches below the surface, with a piece of equal length protruding above the earth. Since the soil is loose and free of clods, these stalks are set out with ease because they are planted as each mound is formed. In other cases, mounds are not prepared, but instead, when the land has been leveled, cleared of weeds, and tilled, these plants are set out at intervals of two or more, close to one another. First, however, the area is cleared and burnt in preparation for planting the cassava, as described in the preceding chapter on corn. A few days later, the cassava begins to sprout, or rather, it takes root, and the planted stalks send forth leaves, and their shoots develop into branches. It is necessary to continue weeding the *conuco* (the land where the cassava is being cultivated) until the plants take over the land. It is beneficial for the cultivated field to be kept clear of weeds at all times. Cassava should always be planted after the new moon and as soon as possible during the days preceding the full moon, but never during the waning of the moon.

This food is in no danger of attack from birds or animals, with the exception of cows, mice, and even horses, since its fruit is a rootstock (*mazorca*), similar to a very large turnip, growing among the large roots of the plant underground. Any man or animal, except the three aforementioned, that consumes these roots without extracting its juice (using a specific type of press), succumbs to death immediately. It is true that on the mainland, there is a type of cassava that is nonpoisonous, but at a glance, it resembles the deadly cassava of this island in terms of its appearance, its branches, fruit, and leaves. The cassava found on this and the neighboring islands of this gulf is, for the most part, the poisonous kind. There is however one type of cassava known as *boniata*, which is similar to that of the mainland, but is not poisonous, and must have undoubtedly come from there. On the mainland, cassava is eaten as a boiled or roasted fruit, given that it is not deadly and also because they do not know how to prepare cassava bread, except in a few parts; and in those places where this is done, it is made from the same kind of cassava used here, in other words, the poisonous kind. As a matter of fact, certain soldiers familiar with these islands have taught the people of the mainland to make bread from the cassava that is not poisonous, so they don't have to be cautious with it. In order to save time, as I have already mentioned, they eat the cassava without making it into bread, but rather boiled and roasted, without squeezing out its juice nor taking the necessary measures to ensure that once the bread is made it will not be poisonous. The *hombres del campo* can always distinguish one from the other. At least it has not been necessary for this to be taught to animals since their natural instinct helps them to protect themselves from such poison (although not all animals), since there are no known cases of death resulting from such cause to any horse, cow, or any other animal of all those brought over from Spain, nor to the countless numbers that have sprung from them. Rather, cattle and mice have been eating cassava every day, as have a number of horses, so insofar as animals are concerned, the cassava does not have the same effect on all of them.

The cassava itself bears a close resemblance to carrots or large turnips from Galicia, sometimes even bigger. In many regions, they usually grow to the size of the calf or thigh of a human. They are covered with a rough skin, dark tawny in color, while some tend to be brown. On the inside, they are extremely white with a texture similar to that of a turnip or chestnut. Large breads called *cazabe* are made from cassava, and this is the staple food of this and many other islands, those inhabited by *los cristianos* as well as those yet to be conquered. It is made in the following manner: After the Indians have removed the skin from the cassava by scraping it with clamshells until every bit is gone, as is done with turnips when they are being prepared for cooking, they grate the peeled cassava on a few rough stones serving as graters for this purpose. The grated cassava is then carried to a very clean place in which they fill a *cibucán*, which is a round-shaped, loosely woven bag, ten or twelve spans in length, as wide as a human leg, approximately, made of soft tree bark, after the style of palm matting. After the *cibucán* has been filled with the grated cassava, a wooden lever and winch are set up, and from this, the *cibucán* is suspended in the air from one of its ends, while weights of large stones are tied to the other end facing the ground. The *cibucán* is then stretched using the winch, raising the stones into the air, suspended in such a manner that the cassava is pressed and all the juice squeezed out, seeping through the woven *cibucán*, and running out onto the ground. The cassava is left in this kind of press until there is not a drop of juice nor must left. This water or liquid is poisonous and is drained out and allowed to run into the ground so as to dispose of it. The squeezed cassava remaining in the *cibucán* is bulky and dry, just as almonds are after they have been pressed out. It is then removed from the *cibucán*. Separate from this area, a hole is made and a fire lit inside of it and a *burén* placed in the center (close to the bottom in order to prevent the fire from spreading). A *burén* is a flat earthenware griddle, as large as a sieve with no sides, below which an intense fire is maintained, without the flame being allowed to get to the griddle, which is firmly kept in position with clay. The *burén* is kept as hot as is necessary, and the cassava (left in the *cibucán* after the juice is extracted), is placed on it as if it were bran or sand, in a round shape. The *burén* is almost completely covered with the cassava, however, a space measuring two fingers is left along the edge, in addition to which, the cassava is piled two fingers high. It is spread out flat and immediately begins to cook. The baker uses a few small, flat sticks as a baker's peel to turn the cassava to cook on the other side. Making cassava bread on the *burén* as described above takes the same time as it does to make an omelette in a frying pan, or perhaps even less. It is then placed in the sun for a day or two in order to dry and turns out to be a very good dish. Wherever there is a large number of people, many *cibucanes* and numerous *burenes* are set up when a large quantity of bread needs to be made.

This bread is good and quite nourishing, plus, it keeps at sea. For everyday people, it is generally made about half a finger thick, and as thin as a wafer and as white as paper for people of rank and standing. This thin cassava bread is called *xauxau*.

Usually, in this city of Santo Domingo, this cassava bread costs a ducat when the price is high and half a peso when it is cheaper. It sometimes reaches as high as its weight in gold (equivalent to 450 *maravedíes*) when it weighs two arrobas,[1] which is fifty pounds of 16 ounces each. This is highly profitable for many people in this region, as this bread is consumed in large quantities.

There are aspects of the cassava plant that are worthy of mention. So much has been said thus far against this backdrop, which could not be more appropriate, that it is fitting to relate the rest. That cassava juice that is extracted from the *cibucán* is so poisonous that one mouthful can kill an elephant or any living animal or man. Nevertheless, if this same deadly juice is boiled two or three times, the Indians can eat it as they would a good stew. However, they do not eat it as it cools. Although it would no longer kill, as it has been boiled, they say it is hard to digest when eaten cold. If this juice is boiled down two-thirds when it is first extracted and placed in the open air for two or three days, it becomes sweet and is used as a sweet liquid, sometimes mixed with their other food items. If, after it has been boiled and left over night, it is boiled yet again and allowed to stand, this juice becomes sour and serves as vinegar or a sour liquid, consumed without any danger whatsoever. This conversion to sweet and sour lies in the boiling, and few of the Indians know how to do this since the old ones are now dead and gone. Furthermore, *los cristianos* have no need for what was described above since there are so many oranges and lemons on the island to provide them with their sour juice. As for the sweet liquid, there is even less need for it as there is sugar in abundance on the island. Consequently, using cassava juice for sweetening purposes and as a sour liquid has since been forgotten. On many occasions, I have witnessed the recently extracted juice of the cassava being consumed as a stew after it has been boiled. I have also seen death resulting from the juice being ingested to quench thirst just as it is extracted, without being boiled, or death caused by simply eating the cassava itself. This is a common occurrence right here and throughout these islands.

Cassava bread keeps for one year and more and has traveled by sea and through these islands and coasts of the mainland. I myself have carried it as far as Spain, as have many others. It is very good bread in these waters and regions, because it can be kept for a long time without going bad, provided that it does not get wet.

This cassava bread known as *cazabe* can be found throughout these islands that I have mentioned. The cassava is not ready to be harvested for bread-making until one year or more after it has been planted. If it is a year and a half to two years old, it is even better and yields more bread. In cases of great necessity, it can be consumed when it is ten months old, but not less.

When this island was inhabited by a large number of Indians, and any one of them wanted to kill himself, he used to eat this cassava just as the rootstock is found and he would be dead within two or three days, or even before that. If however, he drank the juice right after its extraction, there was no time for repentance since this resulted in immediate death. So in order to avoid having to work, as advised by their *cemí* (*diablo*), or for whatever reason they wanted to die, they ended their days by eating this cassava. At times, in seeking to avoid work or having to be servants, they killed themselves together, fifty at a time, by each taking a drink of this juice.

The cassava plantations are a beautiful sight when the cassava is bright and fresh. There are six varieties on this island of Hispaniola. One is called *ipotex*, which bears a fruit like a small apple, each divided into six sections. This variety of cassava is one of the very best. Another is known as *diacanan* and is considered the very best since it yields more bread. The third type of cassava is referred to as *nubaga*. The fourth is called *tubaga*. The fifth is known as *coro*, the sixth and last is called *tabacán*, the branches of which are whiter than any of the others. The individual names of these types of cassava vary in other islands and on the mainland, according to the different languages spoken there.

Corn and cassava bread constitute the principal and essential foods of the Indians. The reader must not ignore the outstanding qualities of cassava that have been highlighted thus far and summarized in the following. Cassava is: food that sustains life; sweet and sour liquids that serve as honey and vinegar; a stew that is eaten and enjoyed by the Indians; firewood from the branches when there is no other; and a potent and deadly poison, as I have previously indicated. Another feature of the cassava bread that comes to mind, one that I was not aware of when this first part of these histories was printed, is that in a certain region of the mainland, excellent wine is made from cassava bread. I will go into more detail in the second part of this *General History*, in Book XXIV, Chapter 3, which will focus on the Huyapari River and the experiences of Captain Diego de Ordaz. Therefore, there are seven noteworthy characteristics of cassava. Let us move on to other matters concerning the agriculture of the Indians.

Part II, Book XXIV, Chapter 3

Focusing on the Huyapari River located in the Gulf of Paria and on what happened to Governor Diego de Ordaz during his time there.

The name Huyapari that was given to this famous river found its origin in *los cristianos* who discovered it during a trip that started in Cubagua with the navigator Juan Bono de Quejo. This name was maintained for a long time until the arrival of Captain Diego de Ordaz. Nevertheless, the natives of the region call it Urinoco. Despite this, the reader must understand that... wherever the name Huyapari appears, we are in fact referring to Urinoco[1]....

It seems to me that it bears some similarity to the Nile, which according to Isidore,[2] inundates and irrigates the land of Egypt and renders it fertile, and which, according to the same author, is occupied by large crocodiles: *solus ex animalibus superiorem maxillam movere dicitur.* However, *Natural History* by Pliny[3] provides further details on the Nile, and indicates that the origin and birth of the Nile is uncertain since it runs through warm, desert areas and through an extremely vast space. It further states that it is inhabited by crocodiles and that at a certain time of the year, it swells and covers Egypt and makes it fertile, and that the flood determines whether the year would be more or less abundant or infertile. The text also indicates that up to the time of Pliny, the greatest flood was recorded at eighteen cubits.

Bear in mind what is said by these authors and listen to me, and you will know what I learned from many witnesses who embarked upon this trip with Ordaz through the Huyapari River. This river rises and falls twenty fathoms; it begins to swell in June and continues to do so until October, when it recedes in like fashion until May. It therefore rises for six months and falls for the rest of the year. Our Spaniards saw it for the first time at the end of December....

Since we previously touched on several characteristics of the Nile when I indicated that the Huyapari is quite similar, and since a few details were given on its rising, before proceeding, I will explain what I understand about this river and the people of the town of Aruacay, where there were nine head *caciques* and one superior referred to as Naricagua, who ordered everyone and was obeyed since he was the *piache* or the high priest. He was the only one among that people who wore a beard. There were two hundred large, round *cabañas* there. When the river rises, it floods the fields on both banks, coming in close proximity to the town, and when it subsides, the people follow its path, along which they sow their crops, until the river returns to its course....

The food prepared includes *cazabe* and wine.... When they want to make wine, they take the "*caninia*" or grated cassava, and allow it to sit for one day without squeezing it, after which it turns sour. The following day, they make *cazabe* or cassava bread. Once the flat cakes have been made, they are dried, washed in water, and placed between annatto leaves, where they are left for two days until they become soft and mouldy, red and sometimes green in color. In this state, they are then dissolved in water, in jars of ten and twelve arrobas that are on hand for this purpose, depending on the desired quantity. The mixture is then left to bubble for three days and ferments the same way that must and grapes do in Spain. After three days, the wine has settled and they drink it clear. It bears a striking resemblance to the new white wine of Castile and lasts up to eight days. . . .

As regards food, I have already explained that there is poisonous cassava and the safe kind; both of which are used to make *cazabe* and that wine mentioned earlier, which is as intoxicating as that of Castile. And if they wish to make it stronger, they add a little ground corn during the fermentation stage. They usually acquire corn in small quantities and it is of great value to them. Their fruits include among others, guava, soursop, cocoa plum, pineapple, hog plum, and prickly pear. . . .

[W]hen there is a death, their custom involves burying the individual in his or her *cabaña*, making a mud coffin that is supported by sticks, and upon this coffin they place the figure of a *diablo* made from the same mud, a calabash with the aforementioned wine, as well as *cazabe*. . . .

Among the other festivities celebrated in Aruacay, there is one main feast celebrated by the Indians in the following manner: all the Indians congregate, men as well as women, some painted in red, others in black and some even using other paints, adorned with all their jewelry and feathers. They make a line with more than one hundred and fifty of the jars used for cassava wine and they place in the middle of them, two larger jars with plates where the handles should be, each one flat and as large as a medium-sized plate. On those flat handles stands one Indian, on both jars, painted and elegant, where he recounts everything he has done in his life and is contemplating doing in terms of battles or personal struggles. . . .

Nota del Editor

En el curso de su "turbulenta pero distinguida"[1] carrera, Gonzalo Fernández de Oviedo y Valdés cruzó el Océano Atlántico doce veces. Durante un intervalo en España, comenzando en 1524, escribió un resumen de su historia del nuevo mundo: *De la natural hystoria de las Indias*, más conocida en ediciones modernas como *Sumario de la natural historia de las Indias*, y publicada en Toledo en 1526. El *Sumario* fue un éxito inmediato, así como también la *Primera parte de la historia natural y general de las Indias, Islas y Tierra firme del mar Océano*, publicada en Sevilla en 1535, año en el que Fernández de Oviedo fuera nombrado comandante del fuerte de Santo Domingo. Para 1544, ya se encontraban en pleno proceso o en discusión, varias traducciones del original en castellano al toscano, francés, alemán, latín, griego, turco y árabe[2].

La segunda parte de *Historia general y natural de las Indias* estuvo lista para su impresión en 1541, pero Fernández de Oviedo no pudo regresar a España hasta 1546, momento en el cual no logró obtener la licencia para su publicación. Se cree que el permiso le fue negado en parte debido a la intervención de Bartolomé de las Casas, ya que la *Historia general y natural de las Indias* discrepaba con la posición de De las Casas sobre el nuevo mundo[3].

Entre 1556 y 1557, Fernández de Oviedo obtuvo permiso para publicar, pero no vivió para ver impresa la segunda parte del manuscrito. También se duda de que Carlos I, quien le encomendara dicho relato, haya visto jamás una copia de la Segunda parte (*Libro XX. De la segunda parte de la general historia de las Indias*), publicada en Valladolid en 1557[4]. Se dieron varias explicaciones para justificar el paréntesis de publicación que se produjera luego de la muerte de Fernández de Oviedo, desde un incendio hasta la reticencia de publicar un relato oficial que criticaba a los clérigos y la conducta de los españoles en las Indias. De todos modos, en 1563, el Rey ordenó al "inquisidor de la Ciudad de Sevilla", a entregar "ciertos libros en su poder 'escritos por la pluma de Gonzalo Fernández de Oviedo, comandante del fuerte de Santo Domingo en la isla de Española, que tratan acerca de las Indias, los que no han sido vistos ni examinados ni autorizados para su publicación' al Consejo de Indias, 'con la debida diligencia y precaución'"[5].

Luego de haber estado en manos de varios dueños y depositarios, en 1775 se encontró una versión casi completa del texto en los archivos del Ministerio de Gracia, Justicia y de las Indias, y el gobierno consideró la posibilidad de publicarlo, pero nunca ocurrió. Tampoco se publicó una edición de principios del siglo XIX a cargo de la Academia de la Historia de Madrid, en parte debido a las revoluciones en la América Hispana. La publicación completa fue llevada a cabo por La Real Academia de la Historia tres siglos después de la muerte de Fernández de Oviedo, bajo la dirección de José Amador de los Ríos. Entre 1851 y 1855 aparecieron publicados, bajo el título *Historia general y natural de las Indias, Islas y Tierra firme del mar Océano*, que comprenden más de 2.400 páginas, más una biografía de cien páginas sobre Fernández de Oviedo escrita por Amador de los Ríos. Este trabajo fue descrito como la "obra más voluminosa escrita desde los tiempos de Adán"[6].

Las publicaciones de 1526, 1535 y 1557 influyeron en las obras de los cronistas y científicos del siglo XVI y de siglos posteriores. El estudio de la *Historia general y natural de las Indias* de Fernández de Oviedo se vio obstaculizado por el hecho de que hubiera muy pocos originales, por el costo de la edición de 1851 así como también por la crítica que por diferentes razones realizaran con posterioridad varios académicos e historiadores de la literatura.

La ausencia de una traducción completa al inglés de la *Historia general y natural de las Indias* constituye un obstáculo serio para el estudio de Fernández de Oviedo. En la actualidad existen varias investigaciones y publicaciones sobre diferentes aspectos de Fernández de Oviedo, pero según entienden los editores, al momento de la publicación de este libro, no existen planes de realizar una traducción completa.

Las páginas siguientes contienen la Primera parte, Libro VII, Capítulo 2 ("Acerca del alimento que los indios llaman cazabe . . . "). Son parte de una traducción al inglés, incompleta e inédita, de la edición española de 1851 de la *Historia general y natural de las Indias, Islas y Tierra firme del mar Océano*, publicada por Amador de los Ríos. Dicha traducción de la *Historia general y natural* estuvo a cargo de Earl Raymond Hewitt (fechas desconocidas) y Theodore Terrones (fechas desconocidas), bajo los auspicios de la Works Progress Administration en 1941. El manuscrito se encuentra en la actualidad en la Bancroft Library de la University of California at Berkeley y nunca se lo menciona entre los materiales citados por Sterling A. Stoudemire en su traducción de 1959 del *Sumario de la natural historia de las Indias*.

Para la presente publicación se encomendó especialmente a la traductora trinitobagüense Aliza Ali la traducción de la Segunda parte, Libro XXIV, Capítulo 3 ("Acerca del río Huyapari, en el Golfo de Paria . . . ") de una edición moderna de la *Historia general y natural de las Indias*, editada por Juan Pérez de Tudela Bueso. En las siguientes páginas figuran aquellos pasajes seleccionados del Capítulo 3 relativos a la yuca. Los textos en español han sido revisados para que reflejen un español actual. Se ha modificado la traducción al inglés de Hewitt/Terrones para que concuerde con la versión revisada en español.

Primera parte, Libro VII, Capítulo 2

Acerca del alimento que los indios llaman cazabe, que es el segundo tipo de alimento que ellos preparan en la isla Española y en otras partes y que, en el presente [1537], también lo hacen los cristianos. Algunos lo utilizan más que el maíz, lo consideran mejor y se sirven más del mismo, el que se hace de una planta llamada yuca.

Hablemos ahora de otro tipo de alimento que los indios hacen de la yuca en la isla Española y en otras islas pobladas por cristianos y también en algún lugar del continente. La planta llamada yuca consta de unos tallos nudosos, algo más altos que un hombre (y otros mucho menores), de un grosor aproximado de dos dedos. El grosor y la altura depende del tipo de tierra, sea fértil o árida, y también de la variedad de planta que se trate. Un tipo de yuca se parece en la hoja al cáñamo, semejante a la palma de la mano de un hombre con los dedos abiertos y extendidos salvo que la hoja de la yuca es más grande y más gruesa que la de cáñamo. Cada hoja tiene siete o nueve puntas o divisiones. El tallo tiene muchos nudos y el vástago es marrón blancuzco, y en algunos casos es casi morado y las hojas son muy verdes y se ven muy bien en el campo siempre y cuando la tierra en la que se cultive esté bien cuidada y libre de malezas.

Existe otro tipo de yuca que no difiere de la anterior en cuanto a las ramas y al fruto pero sí en la hoja ya que aunque cada hoja tiene siete o nueve divisiones, difiere en la forma. Por lo tanto, incluí aquí las formas dibujadas de una y de la otra. De manera que, aunque las hojas tengan algunas similitudes la yuca difiere en sus tipos, ya que unas son más verdes que otras, otras tienen ramas más gruesas, otras varían en la blancura del vástago y muestran diferencias en la corteza, aunque aquí no viene al caso mencionarlas.

Para sembrar esta planta (cualquiera de las que he dicho), se hacen unos montones de tierra, redondos, por orden y en fila como ponen las viñas en el reino de Toledo, y en especial en Madrid que se ponen las cepas a compás. Cada montón tiene ocho o nueve pies en redondo y los bordes de uno tocan con poco margen al otro y dicho montón no es puntiagudo en lo alto sino casi llano y el mismo llega a la rodilla o un poco más. En cada montón ponen seis, ocho, diez o más trozos de la misma planta, vástago o rama de la yuca de manera que cada trozo penetre en la tierra aproximadamente dos pulgadas y quede otro tanto del mismo afuera al descubierto. Como la tierra está apisonada se ponen los tallos de estas plantas con facilidad porque a medida que se van haciendo los montones se van poniendo en ellos las plantas y los trozos de las mismas. Otros no hacen montones, sino una vez limpia la tierra y apisonada, ponen de a trechos estas plantas, de dos en dos o más, cerca unas de otras. Pero primero se tala y quema el monte para plantar la yuca como se dijo en el capítulo precedente respecto del maíz. A los pocos días, nace la yuca o mejor dicho prende, y le salen hojas a aquellos trozos de la planta y sus pimpollos van creciendo en ramas. Es necesario entonces ir desmalezando el *conuco* (tierra de labranza de la yuca) hasta que la planta domine a la hierba y siempre es provechoso que la tierra cultivada esté limpia. Siempre ha de sembrarse después de la luna nueva y lo más rápido que se pueda en los días que preceden a la luna llena pero nunca durante la menguante.

Este alimento no se ve amenazado ni por las aves ni por los animales a excepción de las vacas, los ratones, y los caballos, ya que su fruto es una mazorca en forma de raíz o de nabo muy grande que crece entre las barbas que esta planta echa debajo de la tierra. Cualquier hombre o animal, excepto los tres anteriores, que coma estas raíces con el jugo en la fruta, sin que se exprima el jugo con unas prensas determinadas, muere luego sin remedio alguno. Es cierto que en el continente se encuentra una yuca que no es mortal, la que a la vista se asemeja a la anterior en cuanto a la rama, el fruto y la hoja. Toda la yuca que existe, en su mayor parte, en esta isla y las otras vecinas a este golfo, es de la que mata, aunque también existe una que llaman *boniata*, que es parecida a la del continente, que no mata, y que ciertamente debe haber venido de allá. En el continente se come la yuca como fruta cocida o asada porque allí no es mortífera ni saben hacer alimento de ella sino en pocas partes y en aquellas que sí lo preparan usan la yuca como la de acá, es decir, de la que mata. Es verdad que algunos soldados diestros en estas islas han enseñado en el continente a hacer un alimento con la yuca que no mata, y no se cuidan de ella. Para no perder tiempo pues, la comen como he dicho sin hacerla torta, sino cocida y asada sin exprimirla ni hacer lo necesario para que una vez hecha torta no mate. Entre los hombres del campo siempre se sabe distinguir una de la otra. Al menos no fue necesario enseñarle a los animales ya que su instinto natural los ayuda a protegerse de tal veneno (aunque no todos), ya que no se conoce que se haya muerto caballo, vaca, u otro animal alguno de todos los que trajeron de España, ni los innumerables animales que ellos han engendrado. Más bien, todos los días, vacas, ratones y algunos caballos han comido yuca. De manera que la yuca no tiene igual fuerza en todos los animales.

Las mazorcas de la yuca se asemejan a las zanahorias o los nabos grandes de Galicia y también las hay más grandes, en muchas partes suelen ser del tamaño de la pantorrilla o el muslo de un hombre. Tienen una corteza áspera de un color beige oscuro y algunas tienden al marrón y por dentro son muy blancas y pastosas, como un nabo o una castaña. De estas mazorcas o yuca hacen unas tortas grandes que llaman cazabe y este es el alimento común de éstas y de muchas otras islas, de las que están por conquistar así como de las que están pobladas por cristianos. Se prepara de la siguiente manera. Después de que los indios e indias le han quitado la corteza a la yuca, raspándola hasta que no quede nada como se hace con los nabos antes de echarlos en la olla y una vez sacada la costra con unas conchas de almejas, rallan la yuca en unas piedras ásperas que tienen para este fin. Así, lo que se ha rallado se lleva a un lagar muy limpio donde llenan un *cibucán*, que es un saco largo y redondo de mimbre, hecho de cortezas blandas de árboles, tejido algo flojo y trabajado como una estera de palma, que mide diez o doce palmos de largo y tiene el ancho aproximado de una pierna. Después de que esto se llena con la yuca rallada, se prepara una palanca de madera con su torno y de él se cuelga el *cibucán* por un extremo en lo alto y del otro extremo que cuelga hacia abajo se le atan pesas de piedras grandes y con el torno se estira el *cibucán* y se levanta las piedras en el aire, colgadas de manera tal que se estruja y exprime la yuca y le sale todo el jugo, y la misma se escurre por la tierra por entre las costuras de la tela del *cibucán*. Ahí permanece en este tipo de prensa hasta que no le queda a la yuca ni una gota de jugo o mosto. Este agua o licor es veneno y se vierte y pierde por el suelo cuando así lo desean y lo que queda exprimido de la cibera, adentro en el *cibucán*, se asemeja a muchas almendras secas. Luego toman esto y tienen aparte un *burén* colocado en el fuego en un hueco (que esté bajo para que no se extienda el fuego). Se trata de una cazuela plana de barro, grande como un cedazo, y sin paredes y con mucho fuego debajo sin que la llama alcance a la cazuela que está colocada y fija con barro. Esa plancha que llaman *burén* se calienta cuanto sea necesario y encima echan aquella yuca (que salió exprimida del *cibucán*) como si fuese salvado o arena en torno. Ocupa casi toda la cazuela hasta dos dedos del borde y la yuca llega a dos dedos de alto. La tienden plana. Luego se cuaja y, en lugar de espumadera, con unas tablillas que la hornera tiene para este fin, lo da vuelta para que se cueza de la otra parte. Y una torta de cazabe en el *burén* como la descripta anteriormente se demora lo mismo que una tortilla de huevos en una sartén, o se hace aún más rápido. Después lo tienen un día o dos al sol para que se seque y sale un muy buen alimento. Donde hay mucha gente, ponen muchos *cibucanes* y muchas cazuelas llamadas *burenes* cuando quieren hacer cantidad de ello.

Este alimento es muy bueno. Se mantiene en el mar y para la gente común, lo hacen de medio dedo de grosor y para las personas importantes, tan delgado como obleas y tan blanco como papel, y a esto último llaman *xauxau*.

La carga del cazabe suele valer, en esta ciudad de Santo Domingo, un ducado cuando es caro y medio peso cuando es más barato. Algunas veces llega a un peso de oro (que son cuatrocientos cincuenta maravedíes) cuando la carga es de dos arrobas[1], que son cincuenta libras de dieciséis onzas, y para muchos en esta tierra se obtienen buenas ganancias ya que se consume mucho.

Existen pues características notables de esta planta de yuca y en otro lugar no se podría decir tan a propósito como aquí, donde tanto se ha dicho de este tema y donde es bueno decir lo siguiente. Aquel jugo de la yuca que sale exprimido en el *cibucán* es tan venenoso que un solo trago puede matar a un elefante o a cualquier animal u hombre viviente. Sin embargo, si a este mismo jugo mortal le dan dos o tres hervores los indios lo comen como un buen potaje, pero a medida que se va enfriando lo dejan de comer, porque aunque ya no mata porque está cocido, es de mala digestión comerlo frío. Si cuando este jugo sale lo cuecen hasta que se reduce dos partes y lo ponen al aire libre dos o tres noches, se vuelve dulce, y se lo aprovecha como un licor dulce, mezclándolo con otros manjares, y si después de hervido y dejado en reposo lo vuelven a hervir y reposar, se torna agrio y les sirve como vinagre o licor agrio, sin peligro alguno. Esto de volverse dulce y agrio depende de los cocimientos y pocos indios lo saben hacer porque los viejos ya están muertos y porque los cristianos no lo necesitan, ya que para agrio hay muchas naranjas y limones en la isla que no hay necesidad de lo dicho anteriormente y para licor dulce mucho menos ya que el azúcar abunda en la misma. De manera que se ha olvidado el uso dulce y agrio del jugo de la yuca. Muchas veces he visto lo siguiente: comerlo en potajes después de hervir el jugo salido poco antes de la yuca, apagar la sed bebiéndolo así como queda exprimido sin calentarlo o simplemente comer la misma yuca. Esto se ve aquí y en todas estas islas.

La torta de cazabe se mantiene por un año o más y se lleva por el mar y por todas estas islas y costas del continente. Mucha gente, yo inclusive, la hemos llevado a España y en estos mares y tierras es muy buen alimento, porque se lo aprovecha mucho siempre y cuando no se moje.

En todas estas islas que he mencionado existe este alimento de yuca llamado cazabe. Después de haber pasado un año o más desde su siembra llega el tiempo de recoger el fruto del campo, listo para hacerse cazabe. Si han pasado de año y medio a dos años, es todavía mejor y más rendidor, si hay mucha necesidad se lo puede comer después de los diez meses de sembrado, pero no antes.

Cuando había muchos indios en esta isla y alguno de ellos se quería matar comía de esta yuca, así, como está la mazorca. Y en dos o tres días, o antes, se moría. Pero si tomaba el jugo inmediatamente, no había lugar para el arrepentimiento, porque en seguida se le acababa la vida. Y así, para no trabajar y como aconsejados por su *cemí* (diablo) o porque se les antojaba morir, terminaban sus días comiendo yuca. Ocurrió que algunas veces, para no trabajar ni servir, se mataban muchos juntos, de cincuenta en cincuenta, con sendos tragos de este jugo.

Las plantaciones de yuca en el campo son muy hermosas cuando la misma está linda y fresca. En esta isla Española existen seis tipos diferentes. Una se llama *ipotex*, que tiene un fruto como manzanillas, cada una con seis cuartos. Este tipo es de las mejores. Otra se llama *diacanan* y se considera la mejor de todas porque produce más alimento. La tercer especie de yuca se llama *nubaga*. La cuarta se llama *tubaga*. La quinta se llama *coro*, la sexta y última se llama *tabacán*, cuya rama es más blanca que ninguna otra. Estos nombres varían en las otras islas y en el continente según las diferentes lenguas.

Estos dos cultivos de maíz y del cazabe constituyen el alimento principal y necesario de los indios. El lector no habrá dejado de notar las grandes particularidades que he leído sobre la yuca, que se resumen a continuación: alimento para sustentar la vida, licores dulce y agrio que sirven como miel y vinagre, potaje que comen y gustan los indios, ramas para leña cuando no hubiere otras, y veneno tan potente y malo como he dicho. Otra particularidad que se me ocurre del cazabe que yo no conocía la primera vez que se imprimió esta primera parte de estas historias es que, a su vez, en alguna parte del continente se hace muy buen vino del mismo, como se explicará más detalladamente en la segunda parte de esta *Historia General*, en el libro XXIV, capítulo 3, donde se hablará del río Huyaparí y de los hechos del capitán Diego de Ordaz. Así que son siete las características notables que se encuentran en la yuca. Pasemos a otros temas relacionados con la agricultura de los indios.

Segunda parte, Libro XXIV, Capítulo 3

Acerca del río Huyapari, en el Golfo de Paria y de lo que allí le ocurrió al gobernador Diego de Ordaz.

El nombre Huyapari con el que se denominó a este famoso río tuvo su origen en aquellos cristianos que lo descubrieron partiendo desde Cubagua junto con el navegante Juan Bono de Quejo y dicho río conservó este nombre por mucho tiempo hasta la llegada del capitán Diego de Ordaz. Sin embargo, los nativos de la región lo llaman Urinoco. No obstante esto, el lector deberá entender que como debemos proseguir con el relato de esta gente militar, donde se dijere Huyapari, es en realidad Urinoco[1]. . . .

Me parece que tiene alguna similitud con el Nilo, del que Isidoro[2] dice que inunda y riega la tierra de Egipto y la hace fértil y en el que, como el mismo autor señala, se encuentran grandes cocodrilos: *solus ex animalibus superiorem maxillam movere dicitur*. Pero aquél que desee más detalles acerca del Nilo puede recurrir a la *Historia Natural* de Plinio[3], en la que se dice que el origen y nacimiento del Nilo es incierto porque corre por partes desiertas y calurosas y por un espacio muy vasto y que se crían en él cocodrilos. Agrega que en cierta época del año crece y cubre el Egipto y lo hace fértil y que según sean sus crecientes, el año resultará más o menos abundante o estéril; y dice que su mayor crecida, hasta la época de Plinio, había alcanzado los dieciocho codos.

Recuerde el lector lo que estos autores dicen y escúchenme y sabrán lo que supe de muchos testigos que se encontraban navegando en este viaje de Ordaz por el río de Huyapari. Este crece y disminuye veinte brazas y comienza a crecer en el mes de junio y continúa creciendo hasta el mes de octubre y de ahí en adelante baja de la misma forma hasta el mes de mayo. De manera que crece seis meses y disminuye otros tantos. Nuestros españoles lo vieron por primera vez a fines del mes de diciembre. . . .

Debido a que anteriormente se tocaron algunas particularidades del Nilo cuando dije que el Huyapari se le parece en algunas cosas y que se han dicho algunas cosas sobre sus crecidas, antes de proseguir diré lo que sé de este río y de la gente del pueblo de Aruacay, en el que había nueve *caciques* principales y uno mayor que todos que se llamaba Naricagua, el cual mandaba a todos y era obedecido porque era el *piache* o sacerdote mayor. Era el único entre toda aquella gente que tenía barbas en la cara. La población constaba de doscientas cabañas redondas y grandes. Y cuando el río crece anega los campos de ambas costas hasta muy cerca del pueblo y cuando el río mengua, van tras él sembrando hasta que está en su curso. . . .

Las comidas que preparan incluyen el cazabe y el vino. . . .Cuando quieren hacer vino, toman la *caninia* o masa rallada y la dejan estar durante un día así como está, sin exprimirla, hasta que se fermente. Al día siguiente la transforman en cazabe. Una vez que le dan forma de tortas, las secan y después las bañan en agua y las ponen entre hojas de bijas, donde permanecen durante dos días hasta que se tornan tiernas y mohosas, de color rojo y algunas verdes. Cuando están así, las toman y las deshacen en agua, según la cantidad deseada, en tinajas de diez y doce arrobas que tienen a ese efecto. Así lo dejan hervir durante tres días y se cuece de la misma manera que se cuece el mosto y la uva en España. Pasados los tres días, se asienta y lo beben claro. Esta bebida se asemeja al vino nuevo blanco de Castilla y se mantiene durante ocho días sin echarse a perder. . . .

Y del alimento ya he dicho que tienen yuca de la que mata y de la buena; de ambas hacen cazabe y aquel vino nombrado anteriormente, el que embriaga como el de Castilla. Y si lo quieren hacer más fuerte le echan un poco de maíz molido mientras se cuece. Consiguen poco maíz y lo valoran mucho. Las frutas que tienen son, entre otras, guayabos, guanábanas, hicacos, piñas, jobos y tunas. . . .

Estos indios son idolatros y acostumbran, cuando alguien se muere, a enterrarlo en su cabaña y le hacen una tumba de barro armada sobre palos y encima de ella ponen la figura del diablo, hecha del mismo barro y una calabaza con vino del mencionado, una torta de cazabe. . . .

Entre las otras fiestas que se celebran en Aruacay, los indios tienen una principal que festejan de la siguiente manera. Se juntan todos los indios e indias pintados de rojo y también otros de color negro y otras pinturas, con todas sus joyas y penachos y ponen una fila de más de ciento cincuenta tinajas de vino de cazabe y en medio de todas ellas ponen dos tinajas más grandes que tienen dos asientos por asas, cada uno tan grande como un plato de comida mediano y playo. En aquellas asas planas se pone de pie un indio en ambas tinajas, pintado y elegante y cuenta allí todo lo que ha hecho en su vida y piensa hacer en relación a combates o batallas personales. . . .

PLATE 7
02.01.2000, ARIMA
From CAZABI: *Gift of the Americas*, Volume IV of *12/12*

THE NEW OLD WORLD/*El nuevo viejo mundo*

Archives of Phase I (The Caribbean: 1999–2002)
Archivos de Fase I (El Caribe: 1999–2002)

Marisol Villanueva

Marisol Villanueva was born in 1968 in Isabela, Puerto Rico. She has exhibited her work extensively in Puerto Rico and has shown in group exhibitions in China, Puerto Rico, Spain, and the United States. The archives relating to the Caribbean of THE NEW OLD WORLD/*El nuevo viejo mundo* are the first phase of a long-term curatorial project documenting indigenous communities in the Spanish or formerly Spanish Americas. Materials from Phase I have been the subject of two exhibitions: "THE NEW OLD WORLD: Antilles – Living Beyond the Myth" at the George Gustav Heye Center of the Smithsonian National Museum of the American Indian (also housed online in the NMAI Conexus™ at www.conexus.si.edu/now), and "*CAZABI*: Gift of the Americas" at the Research Institute for the Study of Man (RISM). The other three phases of the project focus on the southwestern United States, Mexico and Central America, and South America. She lives and works in the City of New York.

05.21.2000 CARIDAD DE LOS INDIOS
Marisol Villanueva. Courtesy of James Pepper Henry.

Editor's Note

When Marisol Villanueva began photographically documenting the indigenous communities of the Caribbean in 1999 for Phase I of her curatorial project THE NEW OLD WORLD/*El nuevo viejo mundo*, she hoped that her work would change the way we think about the continuation into the twenty-first century of "New World" ancient cultures. After her first participation in Taíno ceremonies in Puerto Rico, she felt confident that the widespread belief that the autochthonous peoples of the Caribbean had been virtually eliminated by "Old World" forces was incorrect, and that her photographs and other materials would support a different conclusion.

Phase I was spread over four calendar years, 1999 through 2002, and comprised several trips out of the United States of America, where she is now based, to various communities in the insular Caribbean, including her hometown of Isabela on the northwestern coast of Puerto Rico. At the date of this publication, the archives of Phase I include approximately 5000 images, 400 pages of journal entries, and correspondence with the documented communities and with her advisors, including, most notably for this publication, Juan Manuel Delgado.

All of the communities and households documented, irrespective of country or other resources available, shared one vital connection—the cultivation, processing, and use of cassava (or *yuca*, in Spanish). Villanueva was raised on cassava, but before her prefatory research, she knew it only as an everyday food of the people of her country. She was not aware of the physical and spiritual survival of "*la cultura de la yuca*" or of the importance of cassava as a global nutritional resource.

The following text, or "chronicle" (in keeping with the spirit of this publication), is based on the personal journal or *diario* Villanueva kept while documenting the communities. The accompanying photographs include, for the most part, images that were not part of "THE NEW OLD WORLD: Antilles – Living Beyond the Myth" exhibit at the George Gustav Heye Center of the Smithsonian National Museum for the American Indian. Both the text and included images are intended to reveal, through Villanueva's camera lens and experience, how cassava is cultivated, harvested, processed, and prepared as various products for purposes ranging from basic sustenance to sacred offerings.

Villanueva traveled to communities in the Cordillera Central of Puerto Rico, the Cibao region of the Dominican Republic, the Cuban Sierra, Dominica, and Trinidad. She found that the linear and/or spiritual descendants of the autochthonous peoples of the insular Caribbean were perpetuating *la cultura de la yuca*, utilizing implements and cultivation techniques in continuous use for hundreds of years. She observed that cassava is still treated as a sacred food, the communal aspect of growing and enjoying the food has not diminished, and there remains a multigenerational involvement in all stages of cassava production and use.

The text follows the chronology of the *diario*, rather than segmenting her observations by country or community. If a repeat visit was made later on, the dated entry indicates "first visit," "second visit," or "subsequent visit," accordingly. Villanueva began in her own country, Puerto Rico. The text begins in 1999 with two ceremonies in which she participated. The reader is immediately placed in the sacred space of *la cultura de la yuca*, documented in the selected quotations from Juan Manuel Delgado's *Resistencia y supervivencia indígena en Puerto Rico* on page 69. The last entry is from a visit to her hometown, Isabela, in 2002. By then, Villanueva had documented *la cultura de la yuca* in a number of different island nations and understood her own country and its agricultural production in an expanded context.

Also by 2002, she had documented the manufacture and use of woven implements, particularly in Arima, Trinidad, where Ricardo Bharath-Hernandez, Chief of the Santa Rosa Carib Community, and a master weaver, creates these implements for community use and exhibitions of traditional craft.

Villanueva originally wrote her *diario* in Spanish, then translated it into English. While some of the text may be familiar to the reader from the exhibition labels of "THE NEW OLD WORLD: Antilles – Living Beyond the Myth," for the most part, this chronicle is a reconstruction by the editors from information contained in Villanueva's *diario*, and is intended to facilitate comprehension of the images selected for publication. The Spanish translation of the reconstructed texts is by Marisol Villanueva.

With minimal changes, the text and images in this section appear in the previously published exhibition catalog, *MARISOL VILLANUEVA: THE NEW OLD WORLD/El nuevo viejo mundo*, except for the image of the *pastel* (06.16.2002 ISABELA, page 67).

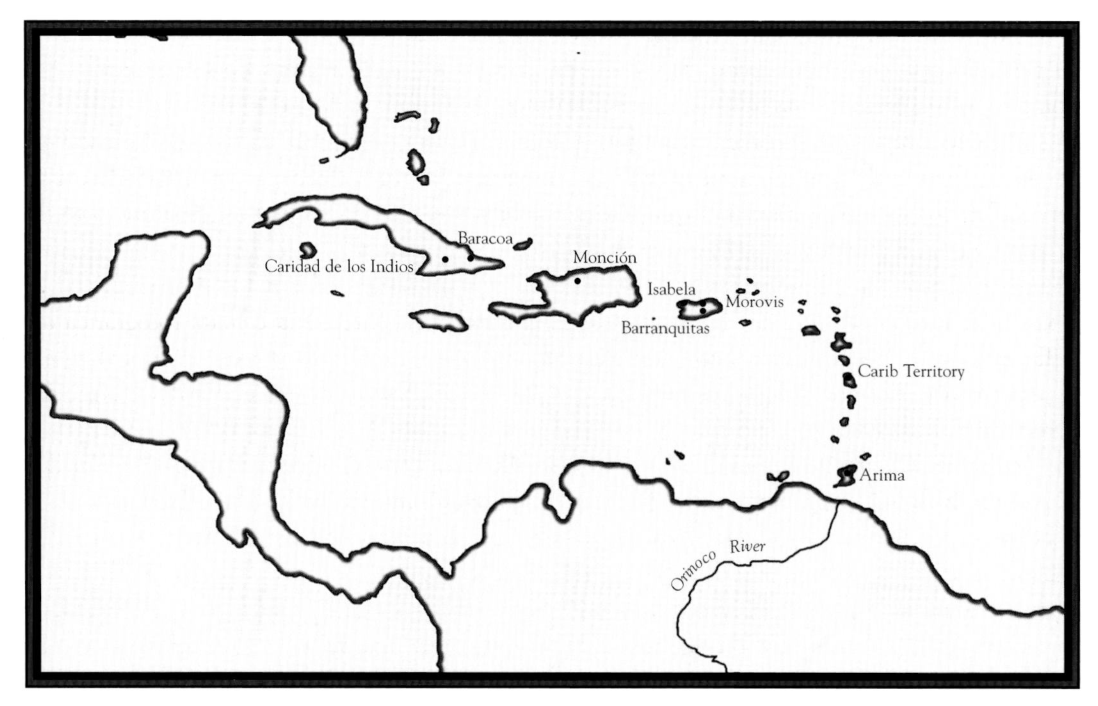

LOCI: Phase I of THE NEW OLD WORLD/*El nuevo viejo mundo* (The Caribbean: 1999–2002)

Nota del Editor

Cuando Marisol Villanueva comenzó, en 1999, a documentar fotográficamente las comunidades indígenas del Caribe para la Fase I de su proyecto THE NEW OLD WORLD/*El nuevo viejo mundo*, ella esperaba que su trabajo cambiara lo que se piensa de la continuidad de las culturas indígenas del "nuevo mundo" en el siglo XXI. Después de su primera participación en ceremonias taínas en Puerto Rico, Villanueva se convenció de que la creencia generalizada de que las fuerzas del "viejo mundo" habían virtualmente erradicado las culturas autóctonas del Caribe era incorrecta y de que su trabajo fotográfico y otros materiales conllevarían a una conclusión diferente.

La Fase I se extendió durante cuatro años, desde 1999 a 2002, y abarcó varios viajes fuera de Estados Unidos, donde Villanueva reside en la actualidad, a varias comunidades del Caribe insular, incluyendo su pueblo natal de Isabela, en la costa noroccidental de Puerto Rico. A la fecha de esta publicación, los archivos de la Fase I incluyen aproximadamente 5.000 imágenes, 400 páginas de anotaciones de su diario y correspondencia con las comunidades documentadas y con sus asesores, entre los que se encuentra Juan Manuel Delgado, de particular relevancia para esta publicación.

Todas las comunidades y hogares documentados, independientemente del país o de los recursos disponibles, comparten una conexión vital—el cultivo, el procesamiento, y el uso de la yuca (o *cassava*, en inglés). La yuca formó parte de la alimentación de Villanueva desde pequeña, pero previo a su investigación preliminar, ella sólo la conocía como parte de la dieta básica de su país. No era consciente cómo "la cultura de la yuca" se había mantenido viva, tanto física como espiritualmente, o de su importancia como recurso alimenticio global.

El texto siguiente o "crónica" (manteniendo el espíritu de esta publicación) está basado en el diario personal que Villanueva escribió mientras documentaba las comunidades. Las fotografías que lo acompañan incluyen, en su mayor parte, imágenes que no formaron parte de la exposición "THE NEW OLD WORLD: Antilles – Living Beyond the Myth", montada por el George Gustav Heye Center del Smithsonian National Museum of the American Indian. El texto y las imágenes incluidas intentan revelar, a través del lente de la cámara fotográfica de Villanueva y de su experiencia, cómo la yuca se cultiva, se cosecha, se procesa, y se prepara en diversos productos, con propósitos que se extienden desde el sustento básico hasta las ofrendas ceremoniales sagradas.

Villanueva viajó a comunidades en la Cordillera Central de Puerto Rico, la región del Cibao en la República Dominicana, la Sierra Cubana, Dominica, y Trinidad. Descubrió que los descendientes lineales y/o espirituales de las culturas autóctonas del Caribe insular perpetuaban la cultura de la yuca usando los instrumentos y las técnicas de cultivo utilizados durante cientos de años. Observó que la yuca todavía es tratada como un alimento sagrado, que el aspecto comunal de producir y de disfrutar del alimento no ha disminuido y que continúa generando un compromiso multigeneracional en todas las etapas de la producción y de su uso.

El texto es fiel a la cronología del diario, en vez de segmentar sus observaciones por país o por comunidades. Por consiguiente, si algún lugar fue visitado más de una vez, la anotación en el diario lo indica como "primera visita", "segunda visita", o "visita subsiguiente". Villanueva comenzó en su propio país, Puerto Rico. El texto se inicia en 1999 con dos ceremonias en las cuales ella participó. El lector se inserta inmediatamente en el espacio sagrado de la cultura de la yuca, documentado en las citas seleccionadas de *Resistencia y supervivencia indígena en Puerto Rico* escrito por Juan Manuel Delgado (página 72). La última anotación es una visita a su ciudad natal de Isabela, en 2002. Para entonces, Villanueva ya había documentado la cultura de la yuca en diversas islas y había logrado comprender a su propio país y su producción agrícola dentro de un contexto más amplio.

También hacia esa fecha, había documentado la fabricación y uso de instrumentos tejidos, particularmente en Arima, Trinidad, donde Ricardo Bharath-Hernandez, el jefe del Santa Rosa Carib Community, y maestro tejedor, crea estos instrumentos para el uso de la comunidad y para exposiciones de arte tradicional.

Villanueva escribió su diario en español, luego lo tradujo al inglés. Mientras que algunas secciones del texto pueden resultar familiares al lector, de las notas de la exposición "THE NEW OLD WORLD: Antilles – Living Beyond the Myth", esta crónica es, en gran parte, una reconstrucción hecha por los redactores, basada en la información contenida en el diario de Villanueva y propone facilitar la comprensión de las imágenes seleccionadas para esta publicación. La traducción al español de estos textos reconstruidos fue realizada por Villanueva.

Esta "crónica" aparece, con cambios mínimos, en el catálogo de exposición, *MARISOL VILLANUEVA: THE NEW OLD WORLD/El nuevo viejo mundo*, publicó anteriormente, con excepción de la imagen en color del pastel (06.16.2002 ISABELA, página 67).

Dates and Locations

07.17.1999 Barranquitas, *Boriquén* (Puerto Rico)

07.21.1999 Barranquitas, *Boriquén* (Puerto Rico)

08.16.1999 Baracoa, *Cuba* (Cuba)

08.27.1999 Monción, *Quisqueya* (Dominican Republic)

09.13.1999 Carib Territory, *Waitukubuli* (Dominica)

01.11.2000 Morovis, *Boriquén* (Puerto Rico)

02.01.2000 Arima, *Cairi* (Trinidad)

02.02.2000 Arima, *Cairi* (Trinidad)

02.21.2000 Monción, *Quisqueya* (Dominican Republic)

02.23.2000 Monción, *Quisqueya* (Dominican Republic)

05.21.2000 Caridad de los Indios, *Cuba* (Cuba)

08.16.2001 Barranquitas, *Boriquén* (Puerto Rico)

06.15.2002 Isabela, *Boriquén* (Puerto Rico)

06.16.2002 Isabela, *Boriquén* (Puerto Rico)

07.17.1999 BARRANQUITAS

07.17.1999 Barranquitas, *Boriquén* (Puerto Rico)

First visit. Last night, I arrived in Puerto Rico to participate in two ceremonies in Barranquitas. I visited Ángel Aníbal López (also known by his Taíno name, Irokahu, or "Man of God") at his home in the densely wooded countryside of the Cordillera Central, where there would be a *Guacar* for his daughter, Yarlier López (known by the nickname, Yari). The *Guacar* ("cave," in Taíno) is a traditional ceremony marking a young woman's first menstruation; the cave represents the womb of *La Madre Tierra* (Mother Earth), from which life springs. The second ceremony was a *Nitainado* for Irokahu, proclaiming him sub-chief to represent the Taíno Nation in Puerto Rico. Gathered at Irokahu's home were his wife, Alida Correa; their four children, including Yari; and his mother, María Esther Berríos. Other participants and guests included Cibanakán, the *cacique* (chief) of the Taíno Nation; a man known almost exclusively by his Taíno name, Baracutey (which means "Solitary"); Nina M. Raffaele Aponte; and four representatives of the Taíno Nation.

Early in the morning, our party went to a clearing in front of a nearby cave, where Yari's *Guacar* was to be performed. Yari's *Guacar*, which welcomes her to womanhood and to a sacred realm of knowledge that belongs only to women, was conducted by Nina. The ceremony began with the women, including Yari's mother, forming a circle and acknowledging the four cardinal points, *La Madre Tierra*, and the center of the circle. Other family members and guests stood apart from the circle. Baracutey then sounded the *guamo* (conch shell), a most sacred Taíno object. Nina invoked blessings from *La Madre Tierra* to assist Yari's entrance to woman-

07.17.1999 BARRANQUITAS
Nina M. Raffaele Aponte

hood and made offerings to *El Gran Misterio* (The Great Mystery). She explained to Yari the responsibilities and duties of womanhood, and guided her in appropriate conduct while menstruating; for example, she should not participate in ceremonies, including the *Guacar*. The offerings of *casabe* (cassava bread), fruit, tobacco, and wildflowers were then placed in the cave, and we all participated in a celebratory dance known as an *areyto*. We then returned to Irokahu's home, where preparations were beginning for his *Nitainado* ceremony, with offerings, including *casabe*, placed on an altar on a ledge outside the house.

07.17.1999 Barranquitas, *Boriquén* (Puerto Rico)

Primera visita. Anoche llegué a Puerto Rico con el propósito de asistir a dos ceremonias en Barranquitas. Fui a visitar a Ángel Aníbal López (también conocido por su nombre taíno, Irokahu, u "Hombre de Dios"), en su domicilio situado en medio de una densa vegetación en la Cordillera Central, donde se iba a realizar la ceremonia del *guacar* para su hija Yarlier López (conocida como Yari). El *guacar* ("cueva" en taíno) es una ceremonia tradicional de la primera menstruación que marca el estado de ser mujer. La cueva representa la matriz de la Madre Tierra, de donde emerge la vida. La segunda ceremonia fue el *nitainado* para Irokahu, en el que sería proclamado sub-*cacique*, representante de la Nación Taína de Puerto Rico. Allí estaban presente

07.17.1999 BARRANQUITAS

su esposa Alida Correa, los cuatro niños (incluyendo a Yari); y su madre, María Esther Berríos. Otros participantes e invitados incluian al *cacique* Cibanakán de la Nación Taína, un hombre conocido exclusivamente por su nombre taíno, Baracutey ("Solitario"), Nina M. Raffaele Aponte y cuatro representantes de la Nación Taína.

Temprano en la mañana fuimos a una cueva, a corta distancia de la casa, para celebrar el *guacar* para Yari. La ceremonia, que fue dirigida por Nina, iniciaría a Yari en la condición de mujer y en una realidad sagrada de carácter mujeril. La ceremonia comenzó con un círculo de mujeres, incluyendo a la mamá de Yari, consagrando los cuatro puntos cardinales, la Madre Tierra y el centro del círculo. Otros miembros e invitados de la familia observaban desde fuera del círculo. Baracutey sonó el *guamo* (concha marina), objeto sagrado taíno. Nina hizo una invocación deseando bendiciones de la Madre Tierra para ayudar a Yari en su condición mujeril, mientras hacia ofrendas al Gran Misterio. Continuó explicándole a Yari sus responsabilidades y deberes como mujer, y le enseñó acerca de la conducta apropiada mientras menstruaba, período durante el cual no debía practicar o participar en ceremonias como la del *guacar*. Las ofrendas de casabe (pan de yuca), fruta, tabaco, y flores silvestres fueron llevadas a la cueva y luego todos participamos en el baile ritual conocido como *areyto*. Después regresamos a la casa de Irokahu, donde se preparaba la ceremonia del *nitainado*, con ofrendas que incluían casabe sobre el altar situado en una repisa fuera de la casa.

07.21.1999 Barranquitas, *Boriquén* (Puerto Rico)

In the morning, Irokahu's mother, María Esther, prepared a delicious ginger tea with milk and butter. While the family got ready for our visit to nearby Cañón San Cristóbal, Irokahu took me to his place of meditation and reflection. It was a cave-like opening formed by the enormous exposed roots and trunk of a *ceiba* (kapok) tree. The *ceiba* is native to Puerto Rico. I believe there are five or six *ceibas* that are hundreds of years old on the island. I followed Irokahu as he tended to his livestock, and before I departed, he showed me part of the land where he cultivates cassava, bananas, and beans, and he explained what each plant yields. He also pointed out wild plants that he and his mother use for medicinal purposes.

07.21.1999 Barranquitas, *Boriquén* (Puerto Rico)

En la mañana, María Esther, la mamá de Irokahu, nos preparó un delicioso té de gengibre con leche y mantequilla. Mientras la familia se arreglaba para nuestra visita al Cañón San Cristóbal, Irokahu me llevó a visitar su lugar de reflexión y meditación: una ceiba con una apertura tipo cueva formada por las raíces y tronco del árbol. La ceiba es original de Puerto Rico. Creo que sólo quedan cinco o seis árboles de tanta antigüedad en toda la isla. Seguí a Irokahu que iba a rescatar el ganado. Antes de irme, Irokahu me mostró parte del terreno en el que siembra yuca, plátanos, y habichuelas (frijoles), y me explicó lo que cada planta proveé. También me mostró las plantas medicinales que él y su madre utilizan como remedios caseros.

08.16.1999 Baracoa, *Cuba* (Cuba)

First visit. In the community of Caridad de los Indios, in the municipality of Manuel Tames in the province of Guantánamo, the *cacique* Francisco Ramírez, known as Panchito, is a spiritual teacher and healer, with a wealth of knowledge about the medicinal plants of Cuba. It is a very difficult and long two-day journey to Baracoa from his village in the Cuban Sierra. His was a rare visit, and I took the opportunity to document Panchito near the Miel River. I was extremely impressed by his intimate knowledge of the plants and their healing benefits. Panchito is wise and gentle, and has a great sense of humor; and we made plans for me to visit his village in the Cuban Sierra at a later time.

08.16.1999 Baracoa, *Cuba* (Cuba)

Primera visita. En la comunidad de Caridad de los Indios, en el municipio de Manuel Tames localizado en la provincia de Guantánamo, al *cacique* Francisco Ramírez se le conoce por el apodo de Panchito. Es el maestro espiritual y curandero, y su conocimiento sobre las plantas medicinales en Cuba es abundante. Aparentemente es muy difícil y largo el viaje de dos días desde la zona que él y una gran comunidad habitan en la Sierra Cubana hasta Baracoa. Aproveché la ocasión de su rara visita para documentarlo en el área del río Miel. Me impresionó su conocimiento profundo sobre la vegetación y sus propiedades curativas. Panchito es un hombre sabio, tierno, y con un gran sentido de humor. Quedamos en planificar para visitar la comunidad en la Sierra Cubana en alguna otra ocasión.

08.27.1999 MONCIÓN

08.27.1999 Monción, *Quisqueya* (Dominican Republic)

First visit. I visited towns in the Cibao region of the Dominican Republic today. I stopped at a home in Monción, where a family was preparing to make *casabe*. Unlike the "sweet" cassava that can be consumed immediately after being boiled, the "bitter" cassava must be processed to extract toxins. If the bitter root is consumed without being processed, the toxic liquid can be dangerous, even though healing properties are attributed to it.

Bitter cassava is processed in the following way in the Dominican Republic: The cassava is peeled, washed, and ground with a hand-grater. It is then thrown into woven sacks where

08.27.1999 MONCIÓN

presses made with large rocks and logs squeeze out the toxic substances. Once the liquid is extracted, the mealy material left over is sifted through a colander to make it into a starchy flour. Then, the cassava is ready for the *burén*, an earthenware griddle still used by some people in this area for making *casabe*.

We then visited another home and where we drank freshly squeezed *güarapo* (sugar cane juice) that had been pressed using the traditional Taíno method. Rice paddies, plantations of cassava, coffee, banana, plantains, aloe, papaya, and other crops occupy a great part of the land here.

08.27.1999 Monción, *Quisqueya* (República Dominicana)

Primera visita. Hoy visité varios pueblos en la región del Cibao en la República Dominicana. Paré en Monción, en la casa de una familia que preparaba casabe. A diferencia de la yuca dulce que al ser hervida puede ser consumida inmediatamente; la yuca amarga necesita ser procesada para extraérsele peligrosas tóxinas. Si es consumida sin ser procesada, el líquido venenoso de la raíz puede ser peligroso, aunque se le atribuye propiedades curativas.

La yuca amarga es procesada de la siguiente manera en la República Dominicana: se pela la yuca, se la lava y se la muele en el molino. Al siguiente, se le echa en sacos tejidos donde las prensas manufacturadas con rocas y troncos de madera exprimen las sustancias tóxicas. Una vez se escurre el líquido tóxico, se filtra la masa para hacer la harina. Inmediatamente, la harina de yuca está lista para la plancha caliente hecha de barro que todavía alguna gente usa por esta zona y que se conoce como *burén*.

Entonces, bebimos *güarapo* (jugo de caña de azúcar) recién exprimido por un hombre en el área con un método tradicional taíno. Gran cantidad del terreno en este país esta ocupado por siembras de arroz, plantaciones de yuca, café, guineo (banana), plátano, sábila, lechoza (papaya) y otras cosechas.

08.27.1999 MONCIÓN

08.27.1999 MONCIÓN

09.13.1999 Carib Territory, *Waitukubuli* (Dominica)

Very early in the morning, I started on my way to see a weaver in the community. When I arrived, one of her daughters and her grandson were washing clothes at the riverside. She explained the process of weaving and showed me how the straw is worked by making a few small baskets. As she wove, hummingbirds and a variety of exotic birds alighted around her, only to depart a moment later.

The weaver and her daughters later prepared a delicious lunch of boiled tubers, including sweet cassava, and fish with coconut sauce.

09.13.1999 Carib Territory, *Waitukubuli* (Dominica)

En la madrugada partí a visitar a una de las tejedoras de paja de la comunidad. Cuando llegué al lugar, me encontré con una de sus hijas y su nieto, lavando ropa a la orilla del río. Ella me mostró el proceso de tejido y preparó unas canastas pequeñas para mostrarme como se trabaja la paja. Mientras tejía, picaflores y otras aves exóticas decendieron sobre ella. Un momento después, desaparecieron.

Luego, la tejedora y sus hijas prepararon un almuerzo delicioso de viandas hervidas, incluyendo yuca, y pescado con salsa de coco.

09.13.1999 CARIB TERRITORY

01.11.2000 Morovis, *Boriquén* (Puerto Rico)

At lunchtime, I visited with Evarista Chéverez Diaz, a potter in Morovis who works in traditional Amerindian forms. Her daughter was cutting breadfruit and cassava to boil. Evarista brewed coffee and then assisted her daughter in preparing the food. I photographed as Evarista and her son and daughter took turns watching over the pot in the kitchen. It was wonderful to witness the deep sense of community within the family, whose members shared cooking duties for even this simple dish.

01.11.2000 Morovis, *Boriquén* (Puerto Rico)

A la hora del almuerzo, visité a Evarista Chéverez Diaz, la alfarera que trabaja con técnicas ancestrales en Morovis. Su hija cortaba pana y yuca para hervir. Evarista coló café y luego ayudó a su hija a preparar la comida. Tomé fotos mientras Evarista, su hijo, y su hija hicieron sus turnos en la cocina. Fue muy bonito atestiguar ese profundo sentido de comunidad dentro de la familia, mientras sus miembros compartían los deberes de la cocina, aunque se trataba de un plato simple.

01.11.2000 MOROVIS
Evarista Chéverez Diaz

02.01.2000 Arima, *Cairi* (Trinidad)

This morning, Ricardo Bharath-Hernandez, the Chief of the Santa Rosa Carib Community, greeted me in the center of the town of Arima. He was weaving a *wareware*, a fan used to control cooking fire. It took him nearly two hours to complete. Later, we began preparations for baking cassava bread. Julie Calderon, Christiana Augustus, and I started peeling, rinsing, and grating cassava. We had a good laugh because they were expecting me to slice my thumb at any given moment, due to my inexperience with their handmade grater. Once we were done—and all my fingers were accounted for—we went to the *ajoupa* (traditional dwelling) next to the Santa Rosa Carib Community Centre. There, we extracted the toxic fluid from the bitter cassava with a long device known in Arima as a *coulevre*. Ricardo had woven this *coulevre*.

The cassava fluid has multiple uses. In Arima, it is used to make *cassareep*, a flavoring and preservative spice used in stews or for seasoning meat. The sediment (a thick, white paste that collects at the bottom of the container when the fluid is allowed to stand) is used to make ironing starch. First, the sediment is left out in the sun to dry; when it becomes a powder, cold water is added, and the mixture is boiled. The sediment can also be boiled and thickened to make glue.

02.01.2000 ARIMA
Julie Calderon (d. 2002) *and Christiana Augustus*

02.01.2000 Arima, *Cairi* (Trinidad)

Esta mañana me recibió Ricardo Bharath-Hernandez, el jefe del Santa Rosa Carib Community, en el centro del pueblo de Arima. Él estaba tejiendo un *wareware*, que es un abanico que se utiliza para avivar la brasa del fuego. El abanico le costó cerca de dos horas de trabajo. Más adelante comenzamos los preparativos para la confección del casabe. Julie Calderon, Christiana Augustus, y yo comenzamos pelando, enjuagando, y rallando la yuca. Nos reímos mucho a cuenta de que ellas estaban esperando que en cualquier momento yo me llevara un pedazo de dedo debido a mi falta de experiencia rallando yuca en ralladores hechos a mano. Una vez que acabamos de pelar y aún con todos los dedos en mano, pasamos a la *ajoupa* (vivienda tradicional) que está al lado del Santa Rosa Carib Community Centre para extraer los tóxicos de la yuca amarga con un aparato largísimo que se conoce en Arima como *coulevre*. Ricardo había tejido este *coulevre*.

El líquido de la yuca tiene usos múltiples. En Arima, se lo utiliza para hacer *casareep*, un condimento y preservativo usado en guisos o para condimentar carne. Se utiliza el sedimento (una espesa pasta blanca que se acumula al fondo del recipiente cuando el líquido se asienta por un rato) para hacer almidón de planchar. Se deja secar el sedimento afuera bajo el sol;

cuando convierte en polvo, se le agrega agua fría y se hierve la mezcla. También se puede hervir y espesar para usar como pegamento.

02.02.2000 Arima, *Cairi* (Trinidad)

Today we made cassava bread again. Julie placed lumps of cassava from the *coulevre* on a handwoven strainer called a *manári*. The loose granular cassava that sifts through is known in Arima as *catebi*. A little later, Julie, Ricardo, and Christiana began cooking the cassava bread on round plates over two small hearths. When the flat, tortilla-like bread is ready, it is placed in the sun for a couple of hours. (Even though cassava bread is usually stored in paper bags or wrapped in plastic, it can sit out for months without any packaging before it goes stale.)

The *catebi* can also be dried in a large pan to make *farine*: the *catebi* is stirred in a hot pan with a stick called an *arabu*. When boiled with water or milk, *farine* is eaten as a porridge. In Trinidad, people also make *farine coocoo*, made of *farine* flour, water, and onion.

02.02.2000 Arima, *Cairi* (Trinidad)

Hoy hicimos casabe otra vez. Julie colocó las masas de yuca provenientes del *coulevre* en un cedazo tejido a mano, que se conoce como *manári*. La harina de yuca se conoce en Arima como *catebi*. Más adelante, Julie, Ricardo, y Christiana comenzaron a cocinar en platos redondos sobre dos fogones pequeños. Cuando el pan, tipo tortilla, está listo, se coloca al sol por un par de horas. (Aunque hoy en día el casabe es almacenado en bolsas de papel o empacado en plástico, puede quedar afuera por meses sin empacar, antes de que se ponga viejo.)

El *catebi* también se puede secar en un sartén grande para hacer *farine*: se agita el *catebi* en un sartén caliente con un palo conocido como *arabu*. Cuando se hierve el *farine* con agua o leche se come como gachas de avena. En Trinidad también preparan *farine coocoo*, hecho con harina de *farine*, agua y cebollas.

02.21.2000 Monción, *Quisqueya* (Dominican Republic)

Subsequent visit. I went to Monción, where I met the mother of Yokastalin Rosario Garcia, also known as "la India." After school, Yokastalin works packing *casabe* with her friend, Sanny Yohani Durán. I also ran into Rosa Maria Gómez weaving an *haigana*, a sack commonly used to carry cassava and *casabe*.

Cassava (bitter and sweet) is widely consumed around the islands of the Caribbean, but I have not seen such widespread use of cassava anywhere else as in the Dominican Republic. You can find it anywhere, in many forms, on this side of Hispaniola.

02.21.2000 Monción, *Quisqueya* (República Dominicana)

Visita subsiguiente. Fuí a Monción donde encontré a la mamá de Yokastalin Rosario Garcia, también conocida como "la India." Al terminar su día escolar, Yokastalin trabaja

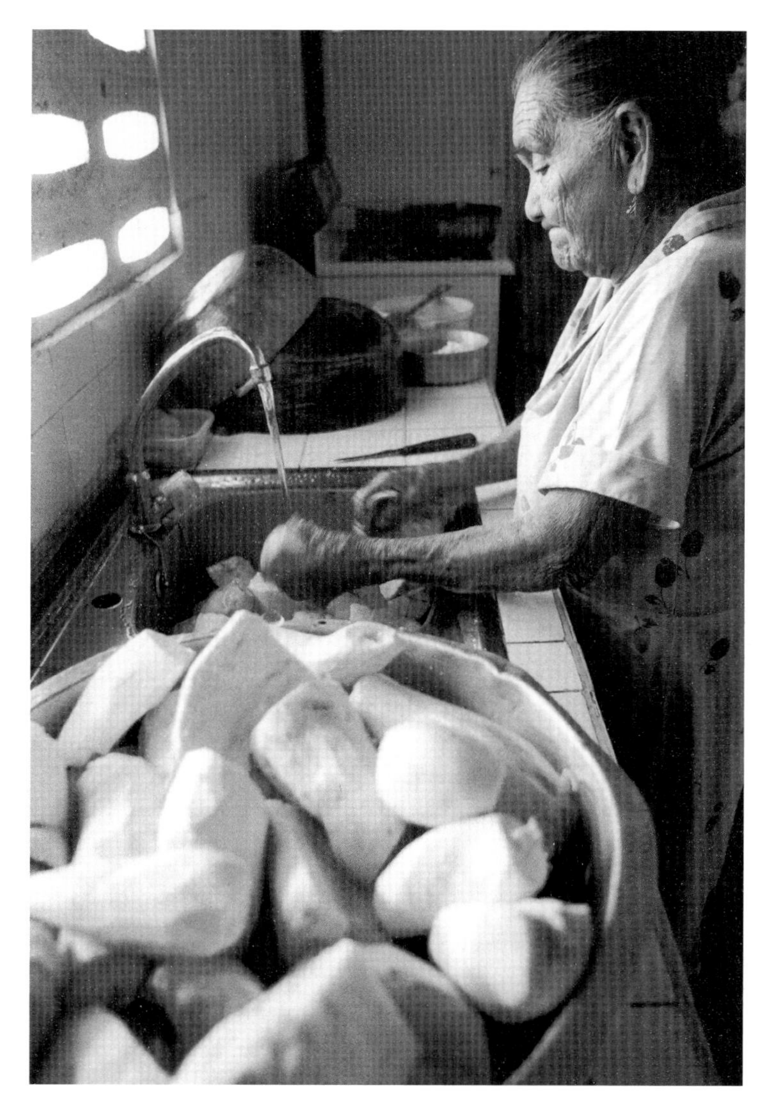

02.01.2000 ARIMA
Julie Calderon

con su amiga Sanny Yohani Durán empacando casabe. Más adelante, me crucé con Rosa Maria Gómez, quien estaba tejiendo una *haigana*, un saco usado comúnmente para cargar yuca y casabe.

La yuca (dulce y amarga) es consumida a través de todas las islas del Caribe, pero no he visto un uso tan extenso como lo que existe en la República Dominicana. En este lado de la Española se puede encontrar casabe en cualquier lugar, y se lo prepara de manera muy variada.

02.01.2000 ARIMA

02.23.2000 Monción, *Quisqueya* (Dominican Republic)

In the morning, I returned to Monción, to the home of Rafaela de Jesus Vargas Blanco. One of her many sons was preparing dough for *panecitos*. He added grated, pressed cassava, anise seeds, and pork skins to the batter; wrapped bits of the mixture in green plantain leaves; and baked them.

02.01.2000 ARIMA
Ricardo Bharath-Hernandez

02.23.2000 Moncíón, *Quisqueya* (República Dominicana)

En la mañana, regresé a Moncíón, a casa de Rafaela de Jesus Vargas Blanco. Uno de sus tantos hijos estaba preparando una masa para hacer *panecitos*. Ví como agregaba yuca rallada, anis, y cuero de cerdo, y envolvía pedacitos en hojas verdes de plátano para hornearlos.

02.01.2000 ARIMA

05.21.2000 Caridad de los Indios, *Cuba* (Cuba)

Second visit. I returned to Cuba to visit *cacique* Panchito. Inside Panchito's home, an altar, which is always present in a corner near the entrance, held offerings including cassava, flowers, and seashells.

Panchito and his extended family prepared a welcome celebration for my travel group. Women filtered coffee and boiled water for cooking, while men outside peeled sweet cassava and got ready to catch a pig for the feast. One member of my group, James Pepper Henry of the Smithsonian National Museum of the American Indian, took photographs of the celebration, including a photograph of me at work. In a matter of minutes, a band began playing *changüi* music, a festive traditional music genre from the Guantánamo province of Cuba.

05.21.2000 Caridad de los Indios, *Cuba* (Cuba)

Segunda visita. Regresé a Cuba para visitar al *cacique* Panchito. Dentro de su casa había un altar sagrado, que siempre está presente en una esquina cerca de la entrada, con ofrendas que incluían yuca, flores, y caracoles.

Panchito y los miembros de su familia extendida prepararon una gran celebración para nuestro grupo de viajeros. Las mujeres colaban el café y hervian agua mientras los hombres pelaban la yuca dulce y se disponian a atrapar un cerdo para el banquete. Uno de los miembros de mi grupo, James Pepper Henry del Smithsonian National Museum of the American Indian, hizo fotos de la celebración, hasta tomó una mía mientras trabajaba. En cuestión de minutos, una banda de músicos comenzó a tocar *changüi*, un género de música festiva tradicional de la provincia de Guantánamo de Cuba.

02.02.2000 ARIMA

02.21.2000 MONCIÓN
Yokastalin Rosario Garcia and Sanny Yohani Durán

08.16.2001 Barranquitas, *Boriquén* (Puerto Rico)

Subsequent visit. I revisited Irokahu and his family in Barranquitas. He was eager for me to try his mother's highly praised *pizza de casabe*, a modern variation on the use of cassava. Around 11:00 AM, I went down to the cassava field with one of the farmworkers. Irokahu warned that we may not be able to get much cassava because the crop was a couple of months younger than it should be for harvesting. Luckily, a few roots were ready.

Soon after, we went to Irokahu's backyard to begin preparing the *pizza de casabe*. Irokahu peeled cassava, while his mother, María Esther Berríos, cleaned a chicken. Alida

Correa, Irokahu's wife, brought us bread and a delicious coffee they grow in the mountains of this region.

María Esther and I went to her house next door to prepare the dough. The process was a shortened version of that followed on the other islands of the Caribbean, but with ordinary kitchen tools. After peeling the cassava, she rinsed it, grated it, and pressed it with her hands in a metal strainer to remove the toxic fluid.

We returned to Irokahu's kitchen, where María Esther made a tomato sauce and heated the cassava dough in a large pan. It took her about fifteen minutes to cook the dough on both sides, then add the sauce and cheese on top of the first side, and let the heat melt the cheese. Irokahu's five children (they had welcomed a fifth child since my first visit for Yari's *Guacar* in 1999) waited eagerly at the table.

08.16.2001 Barranquitas, *Boriquén* (Puerto Rico)

Visita subsiguiente. Regresé a visitar a Irokahu y su familia en Barranquitas. Él estaba deseoso por que probara la tan elogiada pizza de casabe que hace su mamá, y que es una variación moderna en el uso de la yuca. Alrededor de las once de la mañana, bajé a la finca de yuca con uno de los agricultores. Irokahu nos advirtió que posiblemente no ibamos a encontrar mucha yuca

05.21.2000 CARIDAD DE LOS INDIOS

porque la siembra era más joven de lo que necesitaba ser para cosecharla. Afortunadamente, algunas raíces estaban listas.

Justo después, fuimos al patio de la casa para comenzar los preparativos de la pizza de casabe. Irokahu pelaba yuca, mientras su mamá, María Esther Berríos, limpiaba un pollo. Alida Correa, la esposa de Irokahu, nos trajo pan y un café delicioso que ellos cultivan en las montañas de esta región.

María Esther y yo fuimos a su casa, que queda justo al lado de la de Irokahu, para preparar la masa para la pizza. El proceso era una versión acortada del mismo proceso que usan en las otras islas del Caribe, pero con utensilios modernos de cocina. Después de pelar la yuca, María Esther la lavó, la ralló, y la presionó con sus manos en un colador de metal para quitar el líquido tóxico.

Regresamos a la cocina de Irokahu, donde María Esther preparó una salsa de tomate y calentó la masa de yuca en un sartén. Se necesitó unos quince minutos para cocinar la masa por ambos lados. Luego añadió la salsa y queso encima del primer lado y lo dejó cocinar hasta que el queso se derritiera. Los cinco niños de Irokahu (le habían dado la bienvenida a un quinto hijo, desde mi primera visita para la ceremonia del *guacar* para Yari, en 1999) esperaban ansiosos en la mesa.

06.15.2002 Isabela, *Boriquén* (Puerto Rico)

I woke up early to visit a cassava plantation. By the time I arrived, 8:00 AM, plenty of work had already been done, in order to take advantage of the cool morning air. This plantation grows a variety of cassava plants. They can be identified by the leaves, the stalks, and by the color of the pulp. In Isabela, cassava is classified and named by pulp color (white, cream, and yellow). There are many other varieties, including "Jamaica" and "Trinidad," but the two kinds grown on this plantation are "Serallés," named for a family from the south of the island, and "Mantequeira."

The farmer groomed the plant beds, removing weeds and other unwanted items. He cultivates cassava in beach sand, which is very loose and deep. He told me that the cassava plant does not need rich soil to flourish. I asked about some plants that were ten feet tall, and I was surprised when he said they were cassava. They were more than twelve months old. The farmer said harvesting normally begins during the eighth to tenth month after cultivation. The cassava plantations cannot be harvested all at once because as soon as the tuber is harvested, it must be processed within a few days.

The farmer explained that the plants can grow even taller than ten feet, but the root becomes less useful as the plant gets older. He broke this tall plant and proceeded to extract from the ground the largest cassava root I have ever seen in my life. It must have been over two feet long and six inches in diameter. He said this one was almost at the point where he would not be able to sell it. The word for that, here in Isabela, is *jojotes*; it never becomes soft enough to eat, even if it is boiled for hours.

Later in the morning, I visited Nestor A. Ramos Valle at his farm, about a mile from the first farm. Nestor fertilizes his soil. He methodically plants different sections of the land months apart so he can harvest the cassava in stages.

When the soil is ready for cultivation, the process begins with the cutting of six- to ten-inch-long pieces of the stalk, or *cangres*, as they call them locally, from the full-grown cassava plants. The *cangres* are usually cut right before cultivation begins. Nestor's farm also has loose, sandy soil, so the cuttings are placed horizontally in the soil, close to the surface. The farmer then stamps down the soil over the *cangres* with his foot. In other types of terrain, the stalks are planted in mounds.

Nestor plants the cassava according to the traditional Taíno method, during *la luna creciente* (the waxing moon), because it grows underground. Vegetation that yields fruit above ground is planted during *la luna menguante* (the waning moon). Nestor said, "There are two favorable *creciente* moons for cassava cultivation: January and June." However, not everyone plants it in the traditional way; cassava is planted throughout the year by some.

06.15.2002 Isabela, *Boriquén* (Puerto Rico)

Me levanté temprano para visitar una plantación de yuca. Para la hora de mi llegada, a las ocho de la mañana, ya se había hecho mucho trabajo, aprovechando el aire fresco de la mañana. En esta plantación se cultiva una variedad de plantas de yuca. Las plantas pueden ser identificadas por sus hojas, sus tallos, y por el color de la pulpa. En Isabela, la yuca se clasifica por el color de la pulpa (blanca, crema, y amarilla). Hay muchas otras variedades, incluyendo "Jamaica" y "Trinidad", pero las dos clases que se cultivan en esta plantación son "Serrallés", así llamada por una familia del sur de la isla, y "Mantequeira".

El agricultor había preparado las orillas, sacando fuera las hierbas malas y otros objetos indeseados. Él siembra la yuca en arena de playa, que es muy suelta y profunda. Me contó que la planta de yuca no necesita mucho abono para crecer. Yo le pregunté sobre unas plantas que medían más de diez pies de alto, y me sorprendí cuando me dijo que eran plantas de yuca. Estas plantas tenían más de doce meses de sembradas. El labrador me dijo que la cosecha de estas plantas normalmente comienza alrededor del octavo al décimo mes depués del cultivo. La siembra de yuca no puede recogerse toda a una vez, porque una vez cosechado, el tubérculo debe procesarse en pocos días.

El labrador me explicó que estas plantas pueden crecer hasta más altas de diez pies, pero la raíz se vuelve menos útil al cabo de cierto tiempo. Él cortó la planta y procedió a extraer de la tierra la raíz de yuca más grande que he visto en mi vida. Debe haber medido unos dos pies de largo y unas seis pulgadas en diámetro. Me dijo que esa yuca casi estaba al punto de no poder ser vendida. La palabra para describir esto en Isabela es *jojotes*; nunca llega a ser lo suficientemente blanda para poder ser comida, aunque sea hervida por horas.

Más adelante, esa misma mañana, fui a visitar a Nestor A. Ramos Valle en su finca, que queda cerca de una milla de la primera plantación. Nestor fertiliza su suelo, y siembra metódicamente secciones de la tierra en diferentes meses, para así poder cosechar la yuca en diferentes etapas.

Cuando el suelo está listo para cultivar, comienza cortando trozos del tallo de la planta de yuca madura, o *cangres*, como se le conocen localmente, que miden de seis a diez pulgadas de largo. Los *cangres* son usualmente cortados justo antes de la siembra. La finca de Nestor también tiene arena de playa suelta, así que los tallos se colocan horizontalmente en el suelo, cerca de la superficie. El labrador entonces patea el suelo sobre los *cangres* con el pie. En otros tipos de terrenos, se siembran los tallos en montones de tierra.

Nestor siembra la yuca según el método tradicional taíno durante la luna creciente, porque su crecimiento es subterráneo. La vegetación que da fruto sobre la tierra se siembra durante la luna menguante. Nestor dice: "Hay dos lunas crecientes favorables para el cultivo de la yuca: enero y junio". Sin embargo, no son todos que siembran de la manera tradicional; algunos agricultores siembran la yuca durante todo el año.

06.16.2002 Isabela, *Boriquén* (Puerto Rico)

On Sunday morning, I went close to the heart of Isabela to visit a woman who formerly owned a *pastel* factory. In Puerto Rico, *pasteles* are usually made with dense, sticky cassava dough or plantain dough. She showed me the process using cassava. Instead of using a hand grater, she uses a locally made, circular electric grater enclosed in an upside-down pressure cooker. The grated cassava comes out of a spout that leads from the pressure cooker. It operates as well as any modern food processor. After the cassava was ground, it was strained with an old-fashioned coffee filter that looked like a sock. It works very well because the holes are so small, and less pulp is wasted.

I noticed that she had, indeed, saved the toxic liquid to make starch. Some people use the thick, white paste for cooking, she said, by mixing it with water, and adding it to the cassava dough with sugar to make sweet bread.

After she prepared the cassava dough, she cooked the filling for the *pasteles*: meat or chicken, and potatoes cooked in spices. People often add *achiote* (annatto seeds) for color. *Achiote* is a tropical version of saffron and produces a similar reddish-yellow color.

She placed a small amount of cassava dough on a plantain leaf cut into a rectangle roughly the size of a normal sheet of letter paper. The filling is set in the center of the dough, leaving an inch or so of dough free from filling to allow for the *pastel* to be sealed. The plantain leaf is folded to seal the *pastel*, then wrapped with a string made from a palm leaf. Most people use packing string nowadays, but they still cook the *pasteles* in plantain leaves for the flavor.

The *pasteles* can be cooked immediately by boiling them in water for over an hour, or they can be stored uncooked in the refrigerator for an extended period.

06.16.2002 Isabela, *Boriquén* (Puerto Rico)

En la mañana del domingo, fui al corazón del pueblo de Isabela a visitar a una señora que era dueña de una fábrica de pasteles. En Puerto Rico, los pasteles se hacen usualmente con una masa espesa de yuca o de plátano. Ella me mostró el proceso con la masa de yuca. En vez de usar un rallador de mano, ella utiliza un rallador circular eléctrico manufacturado localmente, que incluye una olla a presión al revés. La yuca rallada sale por un conducto que proviene de la olla a presión. Funciona tan bien como cualquier procesador de alimentos moderno. Después de que se muele la yuca, se usa un filtro de café anticuado, que parece una calcetín, para filtrarla. El filtro trabaja bien porque los orificios son muy pequeños y se malgasta menos pulpa.

Me di cuenta de que ella había guardado el líquido tóxico para hacer almidón. Algunas personas comen el almidón, me dijo, mezclando el sedimento blanco espeso con agua, y añadiendole azucar con la masa de yuca para hacer un pan dulce.

Después de preparar la masa de yuca, ella cocinó el relleno para los pasteles: carne o pollo, y papas cocidas en especias. La gente usualmente le añade achiote para darle más color. El achiote es una versión tropical del azafrán y produce un color rojizo-amarillo similar.

Una cantidad pequeña de masa de yuca se deposita en una hoja de plátano cortada en un rectangulo de más o menos el tamaño de una hoja de papel. Se fija el relleno en el centro de la masa, dejando aproximademente una pulgada de masa libre de relleno para poder cerrarlo. Se dobla la hoja de plátano para sellar el pastel, entonces ella ata el pastel con una cuerda hecha de palma. Hoy en diá, la mayoría de la gente utiliza hilo de empacar, pero todavía se continua cocinando los pasteles en hoja de plátano por el sabor.

Se puede cocer los pasteles de inmediato hirviendolos en agua por cerca de una hora, o se puede refrigerarlos por un tiempo extendido.

06.16.2002 ISABELA

PLATE 3
06.15.2002, ISABELA
From CAZABI: *Gift of the Americas*, Volume IV of *12/12*

Selected quotations from

Resistencia y supervivencia indígena en Puerto Rico

Survival of *la cultura de la yuca* in Puerto Rico
Sobrevivencia de la cultura de la yuca en Puerto Rico

a work in publication by

Juan Manuel Delgado

Juan Manuel Delgado has defended the theory of indigenous survival in Puerto Rico for 25 years. He has conducted studies of the oral history pertaining to this issue since 1975. Delgado has published more than 350 works on various historical, sociological, folkloric, and literary topics, as well as poetry, stories, and legends. He has been published in the United States, the Dominican Republic, and Nicaragua. He has also hosted more than 300 conferences on various issues in colleges, schools, and universities throughout Puerto Rico and the United States.

Editor's Note

The following pages contain selected quotations from "*Sobrevivencia de la cultura de la yuca en Puerto Rico*," a chapter from *Resistencia y supervivencia indígena en Puerto Rico*, a book in progress by Puerto Rican scholar Juan Manuel Delgado. The selected quotations are not intended to give a full sense of either the chapter or the complete work, but to provide context and complementary information to other materials in this publication, particularly those of Gonzalo Fernández de Oviedo y Valdés and Marisol Villanueva. Delgado, who met Villanueva in 2001 through people she documented during Phase I of THE NEW OLD WORLD/*El nuevo viejo mundo*, has served in the capacity of an academic adviser to the artist. Delgado's work over the past 25 years documents in part "*la cultura de la yuca*" during the 500-year period between the works of Fernández de Oviedo and Villanueva.

The complete chapter "*Sobrevivencia de la cultura de la yuca en Puerto Rico*" is over 80 typewritten pages, with 5 sections, 48 subsections, and over 200 footnotes. Only the titles of quoted sections and subsections are listed, in a working English translation, as the original contains terms that are difficult to translate into English, if at all. Some footnotes from the original manuscript have been incorporated into the selected quotations, and those that remain appear at the back of this publication.

As the complete chapter is in the process of being edited for publication, the editors of BREAD MADE FROM YUCA have used a rough numbering system for the sections and subsections to give the reader a sense of location in, and the extent of inclusions from, the original, complete chapter. Certain subsections are untitled, being the introductory paragraphs of the chapter or of a specific section. Sections are numbered one through five (1–5). The untitled introductory text to a section is denoted by its section number, followed by a decimal point and a zero (1.0, 2.0, 3.0, etc.). The titled subsections are indicated by their respective section numbers, followed by a decimal point and a one, two, three, etc. (1.1, 1.2, 1.3; 2.1, 2.2, etc.).

If there has been a deletion within a sentence, the absence is indicated by the standard three-dot ellipsis or suspension points. Four-dot ellipses indicate text omitted from the end of a sentence. If an omission has occurred at the start or end of a quoted section, subsection, or paragraph, the ellipses have been omitted to avoid distraction to the reader. Reworded selections or terms added by the editors for clarity are bounded by brackets.

Even in this truncated version, the editors have preserved Delgado's use of "*Boriquén*," "*Borinquen*," and "Puerto Rico" to refer to the names associated with the country during specific time periods. This method helps locate the reader chronologically, in keeping with the theme of this book of selected chronicles. *Boriquén* is used to refer to the society that existed from pre-Columbian times up to the end of the sixteenth century; *Borinquen* is used for the years 1600 to 1812; and Puerto Rico for 1812 to the present.

"*Sobrevivencia de la cultura de la yuca en Puerto Rico*" will be published in its entirety in essay form, with English translation, in 2004.

1 Survival of *la cultura de la yuca* in Puerto Rico

1.0 Puerto Ricans have a particular way of integrating various foods into one meal. By a lengthy process of selection, they have decided that each food should be accompanied by another that complements it. For example, Puerto Ricans eat rice and beans, as well as *pasteles*, but in general, they do not combine rice and beans with *pasteles*. *Pasteles* are eaten with rice and pigeon peas, but never with rice and beans. To describe such combinations, the Puerto Rican uses the expression "that goes with this" or "that does not go with this."

In light of this long-standing tradition, I have observed in various *comida criolla* (Creole food) establishments, that when Puerto Ricans order yuca (also known as cassava), it is combined with fried or stewed fish, since "yuca goes with fish." To a lesser extent, yuca is also combined with stewed, salted cod. This combination is "perfect," especially on the Thursday and Friday of Holy Week. In fact, another age-old Holy Week tradition involves feasting on fish; and there is nothing better than yuca to accompany it. In addition to observing customs, I have interviewed business owners and cooks, and they have insisted that when preparing yuca, the most popular dish is yuca with fish. This led me to pose the following questions: Why is it that five hundred years after the so-called "*Descubrimiento*" (Discovery), Puerto Ricans continue to eat fish and yuca? Why do they prefer these foods on days considered holy? How is it that this combination of foods that "go together" includes two of the most significant in the indigenous diet? After many hours of study and analysis, the answer confirmed the hypothesis. It is a question of another form of indigenous survival, since "*la cultura de la yuca*" in itself, forms part of the survival of that initial culture that resisted and refused to disappear. The following is a presentation of some of my reflections on that survival process.

1.2 *La cultura de la yuca* in the Indigenous Society

Our indigenous people created a production system based on agriculture. . . .[F]or the purpose of this investigation, the most significant aspect of the entire system of indigenous social relations is the irrefutable fact that it involved a culture based on yuca. From that perspective, we refer to their mode of production and social formation as "*la cultura de la yuca*."

When we speak of "*cultura*" we must refer to a totality. This clarification is pertinent since *la cultura* is commonly understood exclusively as artistic or literary expressions. *La cultura* includes all expressions of human activity. From this point of view . . . *la cultura* includes the spiritual expressions of the society, as well as its material base, in other words, economic production and all matters pertaining to said economy. If we use that sociological reality as a point of departure, we must conclude that the indigenous society of *Boriquén*, Puerto Rico today, had as its base *la cultura de la yuca*.

1.4 Yuca: Bread of the Indigenous People

When we refer to "bread" we are not talking about the mixture of flour and water, which, after it has fermented and is baked in the oven, serves as the main food in several countries. I am speaking figuratively, about everything that provides daily sustenance in general; everything that serves as food. This is the context in which the indigenous people saw it. Consequently, yuca and *casabe* (bread made from yuca) were sacred in their eyes.

1.5 Social Coherence and Division of Labor in the Indigenous Society

La cultura de la yuca generated other indirect activities, particularly in the area of craft, which included pottery and basket-making. Skillful artisans, male and female alike, created baskets from *bejucos* (a type of plant native to the Antilles) and the Yarey palm, which were used for storing yuca. They also made bags from cotton threads to store yuca and transport *casabe*. Included among the pottery related to yuca were the *burenes*, earthenware plates used in the cooking of yuca. Another important [class of objects made from clay were artistic sculptures] of *cemíes* (deities) associated with the worship of yuca. There were also stone sculptures that represented the yuca *cemí* or *cemíes*. . . . The entire community was directly or indirectly involved in activities pertaining to yuca. No one escaped and no one could escape the scope of influence of yuca, since it was an essential element of [indigenous] mythology. . . and in light of this, it had to be included in their most significant spiritual activity: the *areyto*.

1.6 The *Areyto* in *la cultura de la yuca*

[The Europeans] saw in the *areyto* a series of activities carried out in Europe, but in a fragmented fashion: religious rituals, dances and important ceremonies, poetry, songs, theater and games. . . . [Activities that were carried out by separate individuals and in separate events in Europe were integrated into one event by the indigenous people in an enormous *batey*. A *batey* is a village square or plaza in which ritual ballgames and other cultural activities were held.] Another significant difference stands out at a glance: the *areyto* was a [communal] activity.

The *areyto* ceremony was one of the aspects that impacted most on the conquistador. Gonzalo Fernández de Oviedo provides a detailed description of this ceremony in his work, *Historia general y natural de las Indias*, Book V, Chapter 1. Fernández de Oviedo witnessed several *areytos* in 1515, including one dedicated to the [legendary *cacica*, female *cacique* or chief] Anacaona [in Hispaniola] by more than 300 women.

When we identify all the objectives and motivations of the indigenous *areyto*, we can draw the conclusion that it was the most significant activity in the Indo-Antillean world. It was an integral social activity in which various arts and disciplines were combined: recreation, entertainment, music, dance, song, poetry, training or education on historical-religious aspects and ceremonies such as weddings, births and other social activities. Although Fernández de Oviedo does not directly state that their myths were transmitted in the *areytos*, it is implied in the expression: "and other things that they want to communicate or make known to the young and old." Without a doubt, part of the *areyto*, or individual *areytos*, were dedicated to religion. In fact, in his work *Relación acerca de las antigüedades de los indios*, Friar Ramón Pané tells us that the "beliefs and idolatry of the Indians and the manner in which they venerate their gods were transmitted orally, from one generation to the next."

1.15 *Casabe* in Oral History

[My extensive research of oral histories indicates] that *casabe* consumption was very common until the mid 1930s. Its decline started from 1940, even though its derivatives continued to be used for medicinal purposes and to acquire starch for ironing clothing. In many parts of Puerto Rico, remote families continued preparing *casabe* until the 1960s, as was the case in the towns of Ciales, Morovis, Utuado, Florida, and San Germán, among others. However, it survived to a greater extent in the northeast section of Puerto Rico.

2 Survival of Yuca Production Techniques

2.5 Survival of Yuca Planting Techniques

The Puerto Rican historian Elsa Gelpí Baíz took an inventory of estates and farms in operation during the period 1525 to 1600. This period is crucial, since it refers to the time when the Spaniards enjoyed greater military control of the country. Even though they had not pacified the entire territory, at least there were periods of relative peace in the periphery of Puerto Rico; currently Viejo San Juan. Indigenous resistance continued, in a type of guerrilla warfare, until 1606, subsequent to which there were further indigenous uprisings.[1] Gelpí's list indicates that all the estates and farms mentioned, 33 altogether, were dedicated to the planting of yuca or to the making of *casabe*; or were devoted exclusively to the preparation of the *conucos* [field where yuca is cultivated]. What is interesting about this statistic is that one of the estates belonged to the indigenous people.[2] The information provided in Gelpí's text strengthens and confirms my working theory that it was yuca and not sugar that formed the basis of the [early] colonial economy [described by Gelpí].

2.6 Yuca Planting and Sexuality

The relationship between yuca planting and sexuality has . . . been expressed in the country's oral tradition. From an early age, I have heard farmers from different regions of the country use the word *montón* (mound) to refer to the female sex organ. I have also heard the word "yuca" used in reference to the male sex organ, as well as expressions such as "they got yuca" and "they want to get yuca," all said in double entendre. These expressions are commonly used in our Caribbean region.

2.7 Yuca Planting and the Spiritual World

The "*cangre*" is what is planted in order to produce yuca. The *cangre* comes from the stem of the *yucubía* [the indigenous name of the plant produced by the yuca] and is cut into pieces to be planted in the earth. The Puerto Rican farmer usually plants the *cangre* in a reclining position, or "*acostada*." When the term is used in this context, it means that the *cangre* is placed horizontally in the ground, or in a slightly leaning position. Puerto Rican archaeologist Jaime Vélez believes that this practice constitutes a survival of a religious nature. He relates the planting of the *cangre* and the preparation of the mound to the burial of a human body.

I cannot conclude without mentioning the matter concerning the spiritual atmosphere created by the farmer in the yuca fields. In my preparations for this project, I interviewed [an agronomist] for his opinion concerning this matter.... His expressions indicate that the yuca field, or the "*conuquito de yuca*," as it was called by the rural population, was a haven of spiritual peace.

[He states:] "The atmosphere in the yuca field is religious. It is a natural affair. The aroma, the tranquility, the ambience in the field is religious. Even if you are not a poet, the atmosphere compels you to make some sort of expression as to the beauty of the field. In that atmosphere, the farmer gave thanks to God for the beauty that was the field and prayed that it bear fruit. They said, '*Dios bendiga esta siembra*' (God bless this field). Even the passersby gave their blessings. I myself, up to today, thank God whenever I am in the field and I speak to the plants."

3 Survival of *Casabe*-Making Techniques

3.6 The Shape of the *Casabe*: A Spiritual Symbol

Based on the information provided by sixteenth-century chroniclers, the indigenous people made *casabe* in different sizes, in addition to varying shapes: circular as well as square. However, according to the oral history, there was a time, many years ago, when moulds were used in order to give the *casabe* a uniform shape. In all geographical and cultural zones of [Puerto Rico], I have heard the story that they previously and currently use the *caldero* (pot) and *sartén* (frying pan) to give the *casabe* its rounded shape.... At the beginning of the twentieth century, *las negras y mulatas* also used several moulds to make the *tortas de burén* [the name used for *casabe* in Loíza].

From where did all this come? It is possible that it all started when the Spanish authorities established colonial regulations governing *casabe* when it was accepted as [a] staple.... It can be deduced from those regulations that *casabe* vendors were compelled to prepare cakes weighing three pounds, two ounces. That stipulation lasted more than two centuries in *Borinquen* and appears recorded as early as 1620. Furthermore, anyone selling *casabe* below the specified weight was immediately fined. Such rigorous legislation must have had a counterpart; a mould made of any material that would guarantee the accurate measurement and weight of the *casabe*, which was... circular. Therefore, five hundred years later, the practice endured in both Loíza and Ciales.

4 The Spiritual Dimension of *la cultura de la yuca*

4.2 San Antonio: The *Casabe Cemí*

[San Antonio is] the *cemí* that is most identified with *casabe*, the main product derived from yuca. *Borinquen* handed down to Puerto Rico a *cemí* exclusively for *casabe*: San Antonio.

San Antonio is one of the saints preferred by Catholics in Puerto Rico.... [There was a widespread,] deep respect, admiration, and affection for San Antonio. With respect to San Antonio, Dr. Marcelino Canino Salgado states:

> San Antonio of Padua: Patron Saint of the towns of Barranquitas, Ceiba, Dorado, Guayama and Isabela. Devotion to him is spread widely throughout the entire country. Petitions for help in finding lost objects are generally presented to him.... His feast day is celebrated on June 13, usually with great pomp, if it falls on a Tuesday. In the churches dedicated to him bread is distributed in his honor.[3]

The information provided by Dr. Canino is very valuable. Nevertheless, it can be further assessed with other information offered by the loose ends of history.

On one occasion, several years ago, I was surprised to see an image of San Antonio in a *conuco* belonging to a farmer. The first thing I thought was that San Antonio was watching over the *conuco*. At that moment, I remembered that the Amerindians used to bury the *cemíes* in the plantations so as to produce large crops of yuca. However, when I asked, [the landowner] denied putting San Antonio in the field to protect his crop, but he could not explain San Antonio's presence. His response was not entirely negative since we lack knowledge of his initial intentions. [A] local countrywoman of the district of San Lorenzo de Morovis...gave me the answer. When I asked her whether or not it was true that San Antonio helped to find boyfriends for young girls, she answered, "No. . . .Whenever my father was going to work in the field, he used to say: *Ay Padre Mío, San Antonio, dame el pan de cada día* (Oh Father, San Antonio; give me my daily bread)." ...[This] elderly woman with strong indigenous features...says that her mother and grandmother were of Amerindian heritage. During our conversation, she explained to me that they only saw San Antonio as the provider of bread. If this is the case, we now understand why her father called on San Antonio when he was going to his *conuco* in search of food. Furthermore, we understand why bread was distributed in his honor in the churches of Puerto Rico.

San Antonio was also seen as a "*bohique*" [shaman]. Throughout the country, verses such as the following were heard:

> Padre San Antonio
> médico divino,
> cura a los enfermos,
> mancos y "tuyíos".
>
> Padre San Antonio
> médico del cielo
> dale la salud
> a todos los enfermos.

Padre San Antonio
médico divino
danos la salud
que te la pedimos.[4]

[*Translator's Note: Some of the rhythm and effect of the original are somewhat altered in the following translation.*

Oh San Antonio
divine doctor,
cure the sick,
the handicapped and the crippled ("tuyíos").

Oh San Antonio
doctor of the heavens
give health to all the sick.

Oh San Antonio
divine doctor
give us the health
we implore.]

4.3 The Religious Foundation of *Casabe*

Casabe was the most sacred possession of the indigenous people. In their eyes, *casabe* was even more sacred than the host of *los católicos*, which as a symbol, was distributed during *la misa de los cristianos* [the Christian Mass]. For *los católicos de la Conquista*, the host was the victim offered to God in sacrifice. That was their symbol. However, materially speaking, it was a piece of unleavened bread, round and thin, that was made for that same sacrifice during the Mass. For the indigenous people, *casabe* was bread that was eaten on a daily basis and it was not necessary to confess their sins to the *bohique* before eating it.

4.4 The *Rosarios Cantaos*: The New *Areyto*

The indigenous *areyto* lived on in the holding of wakes and in the *rosarios cantaos*, or the sung rosaries. At wakes, friends, relatives, and neighbors "kept vigil" for the deceased person all night long. The praying of the rosary was a Catholic tradition but in *Borinquen* it became prayers that were sung. This must have stemmed from the indigenous *areyto* (predominantly sung) and was reinforced by oral poetry sung by the Africans.[5] The Spanish and the wealthy *criollos* recited the rosary, but the rest of the country sang it. That distinction was evident; it was well marked.[6] [In the eighteenth century,] however, apart from that formal difference, there was an essentially radical difference between the two. The Spanish Rosary was solemn; the *rosario cantao* was a *celebración*. . . . [Angel] López Cantos states: "They all sang the rosary, and at the end they feasted on food and especially drink in abundance." At the end of the night, the meeting was a true *celebración* and ended "in games, fandangos, intoxication and entertainment."[7] Therefore, the *rosario cantao* included sung prayers, dances, games, and entertainment like riddles, jokes, gossip, stories, etc. The oral history of the 1970s described those *rosarios cantaos* in same or similar fashion, and was still taking place in that region during the early decades of the twentieth century.

Since the *rosario cantao* was an indigenous *areyto*, it continued to be forbidden by the Church. In 1774, Bishop Manuel Jiménez prohibited it and ordered that a fine of twenty pesos be paid on the first night of the meeting and forty pesos on the second. What is interesting is that for the Church, those celebrations were called *asambleas*, proof of their similarity with the *areyto*.[8] As is the case at all celebrations, during the *rosario cantao*, *casabe* with ginger was eaten, together with chocolate drink or coffee, according to the oral tradition.[9]

4.5 *Juntas*: Another Form of the *Areyto* and Spiritual Survival of Yuca

In studying the development of capitalism in Puerto Rico, we see that during the nineteenth century, there were still several modes of production that co-existed and prevailed in certain regions of the country. . . .The indigenous production method was also being used, which was a type of community system. . . where money was not used and everyone worked in the planting and harvesting of crops, without the factor of salary coming into play. That system was known as "*Juntas*," since the people "joined forces" (from the Spanish verb *juntar*, to join) in order to produce and share.

After the planting was completed, or after the harvesting was done, a *fiesta* was held, which was another kind of *areyto*.

I have been able to learn through oral history that there were *Juntas* for preparing *casabe*. Groups sometimes consisted of as many as twenty. The women grated and cooked while the men were responsible for squeezing the "*cibucán*;" an indigenous word that was still being used at the beginning of the twentieth century. Several *Juntas* for preparing *casabe* were still in operation up to the 1920s, in the district of San Lorenzo de Morovis.

4.6 The Modern Yuca *Areyto*: The Most Diverse Spiritual Expression of the Country

As we have seen thus far, the survival of the indigenous *areyto* was disguised in other cultural expressions like the *Juntas*, devotion to *santos de madera* [wooden saints], and the *rosarios cantaos*. We can however identify the survival of the *yuca areyto* in other folkloric manifestations. . . . This is evident in their songs; in *aguinaldos* (Christmas songs) and in *décimas* (ten-line stanzas); as well as in *poesías*, *refranes*, *adivinanzas*, *canciones de cuna*, *pregones*, *juegos infantiles*, *bailes* (poems, refrains, riddles, lullabies, public announcements, children's games, dances) and other expressions.

The following is one of the many *aguinaldos*.

> Ay seña María
> Ay compae Cortés
> Abrannos la puerta
> Queremos comer
>
> Pasteles o hayacas
> Y un trago de café.

Si me dan pasteles
Dámelos de arroz
De yautía a Pepita
Y de yuca a Margot.
De los pastelitos
Aunque sea uno
Ya sea de masa
Arroz o de yuca.

Yo quiero empanada
De yuca y lechón
Pero antes que nada
Venga mi turrón.[10]

[*Translator's Note: This* aguinaldo *has been summarized in English as the rhythm and effect would be lost in translation.*

The singer and his group of friends are calling out to María and Cortés to open the door and let them in to eat *pasteles* or *hayacas* with a bit of coffee. He continues to sing that he wants a *pastel* made from rice, the one made from *yautía* would be for Pepita, and the cassava *pastel* would be for Margot. He really wants the *pasteles*, even if he can have only one, and it doesn't matter if it is made from dough, rice, or cassava. He also wants cassava empanadas with pork, but first and foremost, he wants to have his *turrón,* or nougat.]

4.9 Yuca as Medicine

The work of Gerónimo Pompa details many of the curative properties of yuca. Using the yuca root, the Amerindians successfully treated external inflammation, including erysipelas. The medicine was prepared in the form of a poultice. The root was first cooked and then crushed. A poultice was also prepared using *casabe*. The cakes were soaked with rum and vinegar. The *casabe* poultice helps fight liver irritation and whitlows [paronychia].

Yuca leaves "placed on the pillow induce sleep; for children, they are placed all over their beds and they relieve mothers from restlessness and insomnia. When placed on the forehead and temples, they provide relief for headaches."

Yuca has other curative properties. The Caribs also prepared medicinal drinks from yuca... [including] a *chicha* from yuca. This drink is made directly from the *casabe*, after it is removed from the *burén*. Once it is moistened, the *casabe* is cut into pieces and left in *tinajas* (vats) with a bit of lukewarm fresh water until it ferments. The result of this process is a wine of superior quality with curative properties. The drinks made from *casabe* are used to treat constipation. Before the *casabe* is removed from the *burén*, sour lemon peel is added; the greener the better.

The starch from the yuca...[is used to treat intestinal complaints]. It is also taken to eliminate inflammation in the gastric mucosa and to treat diarrhea and constipation. The yuca starch is used for angina, for which it is mixed in a glass of water with sugar and lemon. In addition, it alleviates or gets rid of toothache when mixed with rum. While this liquid can be ingested, it can also be applied externally.

The starch was considered an almost magical element since it was used to prevent fainting spells. The yuca powder is mixed with cinnamon powder and the body is dusted with it. The yuca powder is also used to correct vision problems or to treat the eyes when mixed with white wine and the juice extracted from plantain or rue. Many of these remedies have been preserved in the Carib communities in Venezuela and among the rural population in general.[11]

4.11 *La cultura de la yuca* and Indigenous Survival in *Espiritismo Nativo*

The historian Cayetano Coll y Toste was the first to draw attention to the similarities existing between the spiritualist "mediums" and the indigenous *bohiques*. In one of his works written in 1897, Coll y Toste describes the functions of the *bohique* and establishes the following: "When his help was requested for a patient, the '*bohique*' began by '*sugestionar*' *al enfermo* (putting the patient into a trance) and invoking the spirits, as is done today by the spiritualist 'mediums' dedicated to the art of healing."[12]

The *curanderos* (healers) mentioned by Coll y Toste were gradually replaced by doctors trained in modern science. Nevertheless, the rural population of Puerto Rico continued interacting with them as they previously did with the *curandero* and the *bohique*. For example, the indigenous people carried *casabe* for the *cemí*, and also presented the *bohique* with *casabe*. During the 1960s and 1970s, it was still a common practice for those living in the rural areas of Puerto Rico to carry bananas and all sorts of vegetables, fruits, and other gifts for the doctor. This was formerly done with the *curanderos*. There are still many doctors in the towns of the *Altura* (heights or mountains) who are given vegetables by the people when they are consulted. They do not have to pay them with these gifts, but they do it as a tradition.

4.14 Yuca and the World of Caves

Oral history contains [information about historical periods] in which *el espiritismo* (spiritism) was persecuted and suppressed. In towns such as Manatí, Yauco, Vega Baja, Ciales, Florida, and others, I have heard stories of people who were forced to perform their rituals in caves since "it was forbidden by the Government."

Many Puerto Rican caves were known by names such as "La Capilla," "La Iglesia," "El Convento;" and many still carry them today. They are names associated with religious practices of the past. They were clandestine religious practices. . . . I have also known individuals, *espiritistas* identified as being of Amerindian heritage, who say that they continued performing rituals until 1940s by tradition. The practice of carrying *casabe* to the house of the *bohique* was suppressed by the Inquisition, but evidence of these and other suppressed practices can be found just about anywhere. Another practice that was kept alive was the practice of taking the *cemíes* to mountains, caves, rivers, and ravines.

4.15 Yuca and the Heavenly Bodies

This topic always comes up in conversations with older farmers. In the rural areas of Puerto Rico, these older folks preserved the tradition of sowing plants, trees, rootstocks, and any other product, all in keeping with the phases or cycles of the moon. The same applied to the harvesting. Farmers in the area of Comerío planted the "*paslotito*" [*cangre*] of the yuca immediately following the new moon. They could plant from that time but before the end of the last quarter phase. In the district of Palomas de Comerío, there was yuca growing so large that [the roots were referred to] as *yuconas*. Their weight ranged from four to five pounds. However, if they are not planted at that time, it is [the cultivator's] belief that the plant would not produce such *yuconas* and they would be of a poor quality.

In the western region of the country, in the mountains of the interior, there is an oral tradition that says that *tubérculos* (tubers), all plants whose fruit or crops grow below the land, must be sown in the first quarter. This tradition was recorded by a folkloric researcher . . . in the region of Jayuya and Utuado. This information is extremely valuable since it comes from a tradition belonging to one of the regions where indigenous survival is more notable, evident, and proven than elsewhere. For obvious reasons, yuca is linked directly to this agricultural technique, which with all probability, dates back to the indigenous society. . . . Literature focusing on the knowledge of the indigenous people with respect to the position of the stars and their relation to the earthly world has increased. In this dimension, we can also see the relationship between the yuca and its spiritual worldview.

4.16 The *Cangre* of the Yuca and Its Spiritual Dimension

One of the aspects of yuca that has impressed me the most is the respect and admiration felt by farmers and their families for the *cangre*. We have already seen the *cangre* with respect to yuca-sowing techniques and their link to the spiritual world. As regards the word itself, I have always found it interesting that whenever the rural population mentioned the word "*cangre*," their tone of voice changed so as to express certain veneration. The *cangre* was the most sacred part of the yuca. Nowadays, one can still hear expressions like "Ah, the *cangre*! That was the best! Without the *cangre*, there is no yuca!" These expressions still reflect the sacredness of the *cangre*.

The *cangre* is the stalk of the yuca that serves as a seed to reproduce the *yucubía*. . . . The word [*cangre*] was recorded by Augusto Malaret in his *Diccionario de Americanismos* (Dictionary of Americanisms), but it was recorded as a Cuban word. This word was also known in Puerto Rico but its use is more widespread among the older people of the western region and especially the mountains. It can also be heard in Florida and Jayuya. However, the farmers I interviewed in Comerío are not familiar with that word. Some of them know it as "*paslote*." . . . It is interesting to note that the word *cangre* has the same meaning in Colombia, plus it is also referred to as "*cangle*." In the district of Río Grande de Jayuya, a farmer planted *cangres* in mounds of earth and debris; particularly plant debris, which served as fertilizer. The mound resembled those prepared before 1492. Many of the folks occupying the rural areas, when planting the *cangre*, invoked a saint or said: "in the name of the Father, and of the Son, and of the Holy Spirit."

4.17 The Sacred Value of *Casabe*

Puerto Rican historians, anthropologists, and folklorists have done an exemplary job in retrieving information pertaining to various aspects of our culture. When faced with such an extraordinary amount of information, they conclude that it is the legacy of the three "*razas:*" the Amerindians, Africans, and Spanish. Nevertheless, more work is needed to determine which elements belong to which culture.

The rural mothers ... usually chewed the food to be given to their infants to help them eat and digest it. As a matter of fact, boiled yuca was one of the foods they chewed before giving it to their babies.... Much to my surprise, *casabe* was one of the foods chewed by the mothers.... This means that five hundred years after the Spanish invasion, the rural population continued to keep their children on a completely indigenous diet. According to oral tradition, *casabe* could not be thrown away since bread was sacred.

We have learned from the chroniclers that there was a *casabe* known as "*jaujau*," which was a "superior *casabe*, extremely white and soft." The chroniclers also called it "*jabjao*," and it sometimes appears written as "*xabxao*." Even though this word did not survive in Puerto Rico, "it has been preserved in Venezuela, with the same meaning."[13] This type of *casabe* was still being made in Puerto Rico in 1644. During the course of that year, Bishop Damián López de Haro, in a letter in reference to the *casabe* that he was served, states the following: "... [A]nd it is the bread of this land, which they have learnt to eat out of necessity, but for me, I cannot bear to eat it, even though they prepare it in different forms and place it on the table, the best one is the *jaujao*."[14]

This text ... demonstrates that the *jaujao* was still being prepared.... It also indicates that the *jaujao* was intended for the people of rank in the society, especially the representatives of the [Church]. If the *jaujao* was for the *cacique*, the *bohique*, and the *nitainos*, one hundred and fifty years later it was being made for the Governor, the Bishop, priests, and the elite of the colonial aristocracy.

5 Conclusion

5.0 While the *burén* continues to be lit, we recall the noble words of our María Teresa Babín:

> How do you feel now, don *Casabe*, in the midst of so many other exotic geometrically sliced breads, hygienically wrapped up in wax paper? Did the men and children of the past gobble you up with the same desire and speed that other breads are eaten today? Is it painful to see yourself relegated to folklore in your rustic and honored persistence, while those spongy soft breads reign at the table?
>
> Don *Casabe* does not speak, but he observes, making a sporadic appearance in our appetites. He is proclaimed in San Juan by the vendors of Loíza, he unexpectedly emerges amidst the tumult and exhaust of road traffic and continues his rhythm without losing his shape or flavor. Who prepares *casabe* today? From where does it come? In which mysterious *fogón* is it being cooked? My eager palate tastes it with amazement, as if the magic of a protective god transports me to Coayuco, and I am reborn in 1511.[15]

Nota del Editor

Las siguientes páginas contienen citas escogidas especialmente de "Sobrevivencia de la cultura de la yuca en Puerto Rico", un capítulo de *Resistencia y supervivencia indígena en Puerto Rico*, obra aún en preparación del académico puertorriqueño Juan Manuel Delgado. Estas citas no pretenden dar una idea completa sobre todo el capítulo o la obra, si no ofrecer contexto e informaciones que complementen los demás materiales incluidos en la presente publicación, en particular aquellos pertenecientes a Gonzalo Fernández de Oviedo y Valdés y Marisol Villanueva. Delgado, quien conoció a Villanueva en 2001 por intermedio de las personas que ella documentaba durante la Fase I de THE NEW OLD WORLD/*El nuevo viejo mundo*, se ha desempeñado como asesor académico de la artista. El trabajo de Delgado durante los últimos 25 años documenta, en parte, "la cultura de la yuca", en los 500 años que separan las obras de Fernández de Oviedo y de Villanueva.

El capítulo completo "Sobrevivencia de la cultura de la yuca en Puerto Rico" consta de más de 80 páginas escritas a máquina, con 5 secciones, 48 subsecciones y más de 200 notas. En una traducción preliminar al inglés, sólo se han incluído los títulos de las secciones y subsecciones citadas, debido a que el trabajo original en español contiene muchos términos que son difíciles de traducir, cuando tienen traducción.

Debido a que el capítulo completo está siendo editado, los editores de BREAD MADE FROM YUCA han recurrido a un sistema provisorio de numeración de las secciones y subsecciones, a fin de brindar al lector un sentido de la ubicación y extensión de los textos extraídos del capítulo original. Ciertas subsecciones no poseen título, porque se trata de los párrafos que introducen un capítulo o una sección específica. Las secciones se encuentran numeradas del uno al cinco (1–5). El texto introductorio sin título de una sección está indicado por su número de sección, seguido de un punto decimal y un cero (1.0, 2.0, 3.0, etc.). Las subsecciones tituladas se indican con los números de sección respectivos, seguidos de un punto decimal y un uno, dos, tres, etc. (1.1, 1.2, 1.3; 2.1, 2.2, etc.).

Si hubiera una supresión dentro de una frase, la ausencia estará indicada por tres puntos suspensivos o elipsis estándar. La elipsis de cuatro puntos señala que se ha omitido un texto al final de una frase. En el caso de que la omisión ocurra al comienzo o al fin de una sección, subsección citada, o párrafo, se han eliminado las elipsis para no distraer al lector.

Aún en esta versión trunca, los editores han preservado los términos "*Boriquén*", "*Borinquen*", y "Puerto Rico", los que Delgado utiliza para referirse a los nombres asociados con ese país durante períodos específicos en el tiempo. Este método ayuda a ubicar al lector cronológicamente, en sintonía con el tema del presente libro de crónicas selectas. *Boriquén* es utilizado para referirse a la sociedad que existió desde los tiempos precolombinos hasta el final del siglo XVI; *Borinquen* se aplica a los años 1600 al 1812; y Puerto Rico señala el período que se extiende desde 1812 hasta el presente.

"Sobrevivencia de la cultura de la yuca en Puerto Rico" será publicada en su versión completa en forma ensayo, con traducción al inglés, en 2004.

1 Sobrevivencia de la cultura de la yuca en Puerto Rico

1.0 Los puertorriqueños tienen una forma muy particular de integrar alimentos diversos en un mismo plato. Por un largo proceso de selección han decidido que cada alimento debe estar acompañado de otro alimento que le sea armónico. Por ejemplo, un puertorriqueño come arroz y habichuelas [frijoles] y también come pasteles pero por lo general el puertorriqueño no come el arroz y habichuelas con pasteles. Cuando van a comer pasteles lo acompañan con arroz con gandules, pero nunca con arroz y habichuelas. Para describir esas formas de combinaciones el puertorriqueño usa la expresion de "eso pega con esto" o "no pega con esto".

Partiendo de esa larga tradición me puse a observar, en distintos negocios de venta de comida criolla, que el puertorriqueño cuando compra yuca la combina con pescado frito o con pescado guisado porque "la yuca pega con el pescado". También lo puede hacer, en menor grado, con el bacalao guisado. Esta combinación es "la perfecta"; sobre todo durante el jueves y el viernes de la Semana Santa. De hecho, otra tradición centenaria de Semana Santa es comer pescado y nada mejor que la yuca para acompañarlo. Además de observar los hábitos, he entrevistado a los dueños de los negocios, y a los cocineros, y siempre me dicen que cuando preparan la yuca el plato que más se vende es yuca con pescado. Ante ese cuadro me formulé las siguientes preguntas: ¿Por qué los puertorriqueños, después de quinientos años del llamado "Descubrimiento" continúan consumiendo el pescado y la yuca? ¿Por qué lo hacen preferentemente en los días que consideran sagrados? ¿Por qué esta combinación de alimentos, que "pegan" eran dos de los alimentos más importantes de la dieta indígena? Después de muchas horas de estudio y análisis, la respuesta confirmó la hipótesis planteada. Se trata de otra forma de supervivencia indígena porque la cultura de la yuca, en sí misma, es parte de la supervivencia de aquella cultura primaria que se negó y resistió a desaparecer. A continuación presento unas reflexiones sobre ese proceso de sobrevivencia.

1.2 La cultura de la yuca en la sociedad indígena

Nuestros aborígines tenían un sistema de producción que estaba basado en la agricultura. . . . [P]ara efectos de esta investigación, lo más importante de todo el sistema de relaciones sociales indígena es el hecho irrefutable de que se trataba de una cultura basada en la yuca. Desde ese punto de vista es que nos referimos a su modo de producción, y a su formación social, como "la cultura de la yuca".

Cuando hablamos de "cultura" debemos referirnos a una totalidad. La aclaración es pertinente porque a nivel popular se entiende que la cultura se refiere exclusivamente a las manifestaciones artísticas o literarias. La cultura incluye todas las expresiones de la actividad humana. Desde este punto de vista, la cultura incluye las expresiones espirituales de la sociedad, pero también incluye su base material; es decir, la producción económica y todo lo relacionado con dicha economía. Si partimos de esa realidad sociológica, tenemos que concluir que la sociedad indígena de *Boriquén*, el actual Puerto Rico, tenía como base la cultura de la yuca.

1.4 La yuca: El pan de los indígenas

Cuando nos referimos al "pan" no nos referimos a la porción de masa de harina y agua que, después de fermentada y cocida al horno, sirve de principal alimento en algunos países. Me refiero, en un sentido figurado, a todo lo que en general sirve para el sustento diario; a todo lo que sirve de alimento. En ese contexto lo veían los indígenas. Por eso veían a la yuca y al casabe como cosas sagradas.

1.5 Coherencia social y división del trabajo indígena

La cultura de la yuca generaba otras actividades indirectas; sobre todo en el área de las artesanías, como eran los productos de alfarería y cestería. Diestros artesanos y artesanas laboraban en la construcción de canastas fabricadas de bejucos y de la palma del Yarey, que eran utilizadas para cargar la yuca. También se construían bolsas fabricadas de hilos de algodón para cargar la yuca y transportar el casabe. Entre la alfarería relacionada con la yuca estaban los *burenes*, que eran los platos construídos de barro y que eran utilizados para cocinar la yuca. Otra alfarería importante era la que a nivel artístico recreaba los *cemíes* relacionados con el culto a la yuca y las esculturas en piedra, que también representaban al *cemí* o *cemíes* de la yuca. Como podemos observar, las actividades relacionadas con la yuca, directa o indirectamente, involucraban a toda la comunidad. Y nadie escapaba, ni podía escapar, a las esferas de influencia de la yuca porque era un elemento esencial de su mitología. . .y por tal consideración, tenía que estar presente en su actividad espiritual más importante: los *areytos*.

1.6 El *areyto* en la cultura de la yuca

[Los europeos] vieron en el *areyto*, una serie de actividades que en Europa se realizaban pero en forma fragmentada: rituales religiosos, bailes y ceremonias importantes, poesía, canciones, teatro, juegos; en fin, lo que en Europa se realizaba por separado, aquí estaba integrado en una enorme plazoleta o batey. Otra gran diferencia resalta a primera vista. El *areyto* era una actividad de carácter colectivo.

La ceremonia del *areyto* fue una de las cosas que más impactó al conquistador. Gonzalo Fernández de Oviedo le dedica una extensa descripción en su obra, *Historia general y natural de las Indias*, Libro V, Capítulo 1. Fernández de Oviedo fue testigo de varios *areytos* en 1515, incluyendo el *areyto* que más de trescientas mujeres le dedicaron a la [*cacica*] Anacaona.

Cuando identificamos todos los objetivos y motivaciones del *areyto* indígena podemos concluir que se trataba de la actividad más importante del mundo indígena caribeño. Se trataba de una actividad social integral, donde se combinaban distintas artes y disciplinas: recreación, entretenimiento, música, baile, canciones, poesías, la instrucción o educación sobre aspectos histórico-religiosos, y ceremonias como eran las bodas, nacimientos, y otras actividades sociales. Aunque directamente Fernández de Oviedo no dice que en los *areytos* se transmitían sus mitos se infiere en la expresión: "y otras cosas que ellos quieren que a chicos y grandes se comuniquen o sean muy sabidas". Sin lugar a dudas, parte del *areyto*, o *areytos* independientes, eran dedicados al tema de la religión. De hecho, Fray Ramón Pané, en su obra *Relación acerca de las antiguedades de los indios*, nos dice que las "creencias e idolatría de los indios, y cómo veneran a sus dioses eran transmitidas oralmente, de generación en generación".

1.15 El casabe en la historia oral

[Mis investigaciones de historia oral demuestran] que el consumo de casabe se mantuvo en forma generalizada hasta mediados de la década de 1930. A partir de 1940 comienza a declinar aunque se continuaban usando los productos derivados con fines medicinales y para obtener almidón para planchar la ropa. En muchos lugares de Puerto Rico, familias aisladas lo continuaban preparando hasta la década del 1960, como es el caso de los pueblos de Ciales, Morovis, Utuado, Florida, San Germán, entre otros. La sobrevivencia mayor se dió en el noreste de Puerto Rico.

2 Sobrevivencia en las técnicas de la producción de yuca

2.5 Sobrevivencia de las técnicas indígenas de sembrar la yuca

La historiadora puertorriqueña Elsa Gelpí Baiz realizó un inventario de las haciendas y estancias que estuvieron en producción durante los años 1525 al 1600. Este período es importante porque se refiere al espacio de tiempo en que los españoles habían logrado un mayor control militar del país. Aunque no habían logrado pacificar todo el territorio, por lo menos tenían unos períodos de relativa tranquilidad en la periferia de Puerto Rico, el actual Viejo San Juan. La resistencia indígena se prolongó, en su modalidad de guerra de guerrillas, hasta el 1606. Y con fecha posterior ocurrieron otros levantamientos indígenas[1]. El inventario de Gelpí demuestra que todas las haciendas y estancias mencionadas, 33 en total, se dedicaban a la siembra de yuca, o producción de casabe, o se dedicaban exclusivamente a la producción de *conucos* [siembra para el cultivo de yuca]. Lo interesante de esta estadística es que una de las haciendas pertenecía a los indios[2]. Los datos suministrados en la obra de Gelpí refuerzan y confirman mi apreciación anterior, a los efectos de que la yuca, y no el azúcar, era la base de la economía colonial que se intentaba montar.

2.6 La siembra de yuca y la sexualidad

La relación entre la siembra de la yuca y la sexualidad . . . ha quedado manifestado en las tradiciones orales del país. Desde temprana edad escuché a campesinos de distintas zonas del país utilizar la voz "montón" para referirse al sexo de la mujer. Pero también escuché la voz "yuca" para referirse al sexo del hombre. Y también expresiones que se usaban como "le dieron yuca", o "quiere que le den yuca", dichas en doble sentido. Estas expresiones son muy comunes en nuestro entorno caribeño.

2.7 La siembra de la yuca y el mundo espiritual

El "*cangre*" es lo que se siembra para reproducir la yuca. El *cangre* proviene del mismo tallo de la "*yucubía*" [nombre indígena de la planta que produce la yuca] y es cortado en pedazos para sembrarlos en la tierra. El campesino puertorriqueño tiene la costumbre de sembrar el *cangre* en forma "acostada". Cuando el agricultor dice que los siembra "acostados" se refiere a que el *cangre* es puesto horizontalmente en la tierra, o en posición un poco recostada. El arqueólogo puertorriqueño Jaime Vélez piensa que dicha práctica es una supervivencia de carácter religioso. Relaciona la siembra del *cangre* y del montón (antes "montones") con el entierro del cuerpo humano.

No puedo terminar estos apuntes sin mencionar el tema del ambiente espiritual que crea en el agricultor las siembras de yuca. Cuando me disponía a entregar este trabajo entrevisté a un agrónomo para obtener de él su opinión respecto a este tema. . . .Sus expresiones demuestran que la siembra de yuca; o el "*conuquito* de yuca", como le decía el jíbaro [campesino]; era un remanso de paz espiritual. Él nos dice:

"El ambiente en la siembra de yuca es religioso. Es una cuestión natural. El aroma, la quietud, el ambiente de la siembra, es religioso. Aunque no seas poeta el ambiente te obliga a hacer cualquier tipo de expresión sobre lo bonita que se ve la siembra. Y en ese ambiente el agricultor le daba las gracias a Dios por lo hermosa que estaba la siembra y le pedía que esa siembra diera frutos. Decían, 'Dios bendiga esta siembra'. Y hasta la gente que pasaba por el lado de la siembra le echaba bendiciones. Hasta yo mismo, todavía en estos tiempos, le doy gracias a Dios cuando estoy en la siembra y le digo mis cositas a la planta".

3 Sobrevivencia en las técnicas de elaborar el casabe

3.6 La forma del casabe: Un evidente signo espiritual

De acuerdo a los datos de los cronistas del siglo XVI, los indígenas preparaban el casabe de distintos tamaños. También lo elaboraban de distintas formas: preparaban casabe en forma circular y también preparaban *casabe* en forma cuadrada. Pero de acuerdo a la historia oral se perfila que hubo un tiempo muy pretérito y lejano en que se utilizaban unos marcos para darle una forma uniforme al *casabe*. En todas las zonas geográficas y culturales del país he escuchado la historia oral referente a que usaban y usan el caldero y el sartén para darle una forma circular a la torta de *casabe*. . . . En Loíza, las negras y mulatas de principios de siglo también usaban unos marcos para hacer las tortas de *burén* [el nombre para casabe en Loíza].

¿De dónde viene todo esto? Es posible que el punto de partida fuese la reglamentación colonial que las autoridades españolas crearon para el casabe al tener que aceptarlo como el pan de América.

Y de esa reglamentación se desprende que obligaban a los vendedores de cazabe a elabo-rar tortas de tres libras y dos onzas. Ese requisito duró más de dos siglos en Puerto Rico y su reglamentación aparece por escrito en fecha tan temprana a 1620. Más aún, toda persona que vendiere tortas de casabe, por debajo de ese peso, era multada inmediatamente. Una legislación tan rigurosa y tan precisa tenía que tener un contraparte; un marco de cualquier material que garantizara la medida y peso de la torta de casabe, la que evidentemente era circular. De modo que, quinientos años después, la práctica quedaba como un reducto, lo mismo en Loíza que en Ciales.

4 La dimensión espiritual de la cultura de la yuca

4.2 San Antonio: El *cemí* del casabe

[San Antonio] es el *cemí* identificado con el casabe, es decir, con el principal producto derivado de la yuca. Borinquen legó a Puerto Rico un *cemí* exclusivamente para el casabe: San Antonio.

San Antonio es uno de los santos preferidos de los católicos en Puerto Rico . . . . [Toda la gente] sentía un gran respeto, admiración y simpatía por San Antonio. Sobre San Antonio nos dice el Dr. Marcelino Canino Salgado lo siguiente:

> San Antonio de Padua: Patrón de los pueblos de Barranquitas, Ceiba, Dorado, Guayama e Isabela. Su culto se halla muy difundido por todo el país. Suele invocársele para encontrar objetos perdidos. . . . Su fiesta se celebra el día 13 de junio y suele realizarse con mayor pompa, si cae martes. En las iglesias de su advocación se reparten panes en su honor[3].

La información que ofrece el Dr. Canino es muy valiosa. Sin embargo, se puede complementar con otra información que ofrecen los cabos sueltos de la historia.

En una ocasion, hace varios años, quedé sorprendido cuando ví un San Antonio en un *conuco* de un campesino. Lo primero que pensé era que San Antonio estaba cuidando el *conuco*. En ese instante recordé que los indios enterraban los cemíes en las plantaciones para producir mucha yuca. Sin embargo, al preguntarle al pequeño propietario negó que pusiera a San Antonio en la tala para proteger su cosecha pero no pudo precisar porque la presencia de San Antonio. Esa respuesta no es totalmente negativa porque no sabemos sus intenciones originales. De todas maneras, una campesina natural del barrio San Lorenzo de Morovis. . . me dió la respuesta. Al preguntarle que si era verdad que San Antonio servía para buscarle novio a las muchachas o jóvenes, me contestó: "No. . . . Cuando papá iba a abiar decía: Ay Padre Mío, San Antonio; dame el pan de cada día". . . . [Esta] anciana de fuerte fenotipo indígena . . . dice que su madre y abuela eran de raza india. Al conversar con ella, me explicó que ellos solamente visualizaban a San Antonio como el proveedor del pan. Si eso es así, entendemos porque su padre convocaba a San Antonio cuando iba a buscar los alimentos a su *conuco*. Y entendemos porque se repartía pan en su honor en las iglesias de Puerto Rico.

San Antonio también era visto como un "*bohique*", un médico curandero. Por todo el país se escuchaban coplas como estas:

Padre San Antonio
médico divino,
cura a los enfermos,
mancos y "tuyíos".

Padre San Antonio
médico del cielo
dale la salud
a todos los enfermos.

Padre San Antonio
médico divino
danos la salud
que te la pedimos[4].

4.3 La base religiosa del casabe

El casabe era lo más sagrado que tenía el indígena. Para ellos el casabe era más sagrado que la hostia católica, que como símbolo se ofrecía en la misa de los cristianos. Para los católicos de la Conquista la hostia era la víctima que se ofrece a Dios en sacrificio. Ese era su plano como símbolo. Pero materialmente era una hoja de pan ázimo, redonda y delgada, que se hacía para ese mismo sacrificio en la misa. Para el indígena el casabe era un pan que se consumía todos los días y no era necesario confesar los pecados ante el *bohique* antes de consumirlo.

4.4 Los rosarios cantaos: El nuevo *areyto*

Ese *areyto* indígena siempre continuó con vida en las celebraciones de los velorios y los rosarios cantaos. En los velorios los familiares y amigos y vecinos se quedaban "velando" al difunto durante toda la noche. El rezo de los rosarios era una tradición católica, pero en Borinquen se transformaron en unos rezos cantados. Esa modalidad tiene que haber arrancado del *areyto* indígena (predominantemente cantado) y fortalecido por las poesías orales cantadas de los africanos[5]. Los españoles y criollos pudientes realizaban rosarios, pero el resto del país los realizaban cantados. Esa division era tajante, bien marcada[6]. Pero más que esa diferencia formal había una diferencia esencialmente radical entre ambos. El rosario español era solemne; el rosario cantao era una fiesta, un verdadero *areyto* indígena. Dice [Angel] López Cantos: "Todos cantaban el rosario, y al terminar los agasajaban con comida y sobre todo con abundante bebida". Al final de la noche la reunión era una verdadera fiesta y terminaba "en juegos, fandangos, embriaguez y diversiones"[7]. De modo que en el rosario cantao habían rezos cantados, bailes, juegos, entretenimientos como adivinanzas, chistes, cuentos, historias, etc. . . . La historia oral de la década de 1970 describía esos rosarios cantaos en igual o forma parecida, y todavía se celebraban en ese ambiente en las primeras décadas del siglo XX.

El rosario cantao, por ser un *areyto* nativo, continuó siendo prohibido por la Iglesia. En 1774 el Obispo Fray Manuel Jiménez los prohibió y ordenó el pago de una multa de veinte pesos la primera noche que se reunieran y cuarenta la segunda noche. Resulta interesante que para efectos de la Iglesia esas fiestas eran "asambleas", lo que evidencia su similitud con el *areyto*[8]. Como en todas las fiestas, en el rosario cantao se consumía el casabe tomado con gengibre, con chocolate o con café, de acuerdo a la tradición oral[9].

4.5 Las Juntas: Otra forma de *areyto* y sobrevivencia espiritual de la yuca

Al estudiar el desarrollo del capitalismo en Puerto Rico vemos que todavía en el siglo XIX había varios modos de producción que coexistían y eran dominantes en ciertas regiones del país. . . . Pero también funcionaba el modo de producción indígena, una especie de sistema comunitario, o de comunismo primitivo, donde no usaban la moneda y todos trabajaban en la siembra y en la cosecha sin entrar el factor del salario. Este sistema era conocido como el de las "Juntas", porque se "juntaba" la gente para producir y compartir.

Después que terminaban la siembra; o después de que terminaban la cosecha; hacían una fiesta que era otra especie de *areyto*.

Sin embargo, he podido evidenciar, a través de la historia oral, Juntas para sacar el casabe. Los grupos a veces llegaban hasta 20. Las mujeres rallaban y cocinaban y los hombres se dedicaban a esprimir el "*cibucán*"; palabra indígena que todavía se usaba a principios del siglo XX. Algunas Juntas para preparar casabe estuvieron funcionando hasta la década de 1920 en el barrio San Lorenzo de Morovis.

4.6 El *areyto* moderno de la yuca: La más diversa expresión espiritual del país

Como hemos visto hasta aquí, el *areyto* indígena tuvo una sobrevivencia disfrazada en otras manifestaciones culturales como las Juntas, la veneración de los santos de madera, y los rosarios cantaos. Pero la supervivencia del *areyto* de la yuca la podemos encontrar en otras manifestaciones folklóricas. . . . La podemos encontrar en sus cantares; en aguinaldos y décimas; también en poesías, refranes, adivinanzas, canciones de cuna, pregones, juegos infantiles, bailes, y otras expresiones.

Entre la gran variedad de aguinaldos veamos esta muestra:

Ay seña María
Ay compae Cortés
Abrannos la puerta
Queremos comer
Pasteles o hayacas
Y un trago de café.

Si me dan pasteles
Dámelos de arroz
De yautía a Pepita
Y de yuca a Margot.
De los pastelitos
Aunque sea uno
Ya sea de masa
Arroz o de yuca.

Yo quiero empanada
De yuca y lechón
Pero antes que nada
Venga mi turrón[10].

4.9 La yuca como fármaco

La obra de Gerónimo Pompa recoge muchas de las propiedades curativas de la yuca. Con la raíz de la yuca, los indios curaban las inflamaciones exteriores, incluyendo las irisipelas. Este fármaco era preparado en forma de cataplasma. Primero cocinaban la raíz y luego la machacaban. Pero también preparaban cataplasmas de casabe. Las tortas eran humedecidas con aguardiente de caña y vinagre. El cataplasma de casabe sirve para combatir las irritaciones del hígado y en los panadizos.

Las hojas de la yuca "puestas en la almohada provocan el sueño. A los niños se les ponen en toda la extension de sus camitas y liberan a las madres de muchos desvelos e insomnios. Colocadas en la frente y sienes, hacen desaparecer los dolores de cabeza".

La yuca tiene otras propiedades curativas. Los caribes también preparaban bebidas medicinales provenientes de la yuca...[incluyendo] una chicha de la yuca. Esta bebida se obtiene directamente del casabe; después que sale del *burén*. Después de ser humedecido, el casabe se corta en pedazos y se deja en tinajas con un poco de agua dulce tibia hasta que fermente. De ese proceso surge un vino de buena calidad que tiene propiedades curativas. Los atoles de casabe se usan para combatir los pujos (estreñimiento). Antes de que el casabe salga del *burén* se le aplica una cáscara de limón agrio; entre más verde mejor.

El almidón de la yuca sirve para lavativas en las afecciones intestinales. También se toma para eliminar las inflamaciones de la mucosa gástrica y también para eliminar las diarreas y pujos. El almidón de la yuca se utiliza para anginas, para lo cual solamente se mezcla en un vaso de agua con azúcar y limón. También alivia o elimina los dolores de muela cuando se mezcla con aguardiente. Además de usar buches de este líquido puede aplicarse a nivel externo.

El almidón era considerado como una propiedad casi mágica pues se utilizaba para detener los síncopes. El polvo de la yuca se mezcla con el polvo de la canela y se procede a empolvar el cuerpo. También el polvo de la yuca corrige problemas de la vista o para curar los ojos cuando se mezcla con vino blanco y el zumo del llantén o de la ruda. Muchos de estos remedios han logrado sobrevivir en las comunidades de caribes en Venezuela y entre los campesinos en general[11].

4.11 La cultura de la yuca y la supervivencia indígena en el espiritismo nativo

El historiador Cayetano Coll y Toste fue el primero en llamar la atención sobre las semejanzas existentes entre los "*mediums*" espiritistas y los "*bohiques*" indígenas. En un trabajo escrito en 1897, Coll y Toste describe las funciones del *bohique* y establece lo siguiente: "Cuando sus auxilios eran solicitados para un paciente, empezaba el '*bohique*' por 'sugestionar' al enfermo, haciendo una invocación a los espíritus, como lo hacen hoy los '*mediums*' espiritistas que se dedican al arte de curar"[12].

Los curanderos que menciona Coll y Toste fueron sustituídos gradualmente por médicos formados por la ciencia moderna. Sin embargo, los campesinos de Puerto Rico continuaron relacionándose con ellos de la misma forma que lo hacían con el curandero y antes con el *bohique*. Por ejemplo, los indígenas le llevaban casabe al *cemí*, pero también le llevaban casabe al *bohique*. Todavía en las décadas de 1960 y 1970 era muy común que los campesinos de Puerto Rico le llevaran plátanos y todo tipo de verduras, frutas y otros obsequios al médico. Antiguamente, lo hacían con los curanderos. Todavía muchos médicos de los pueblos de la "Altura" reciben verduras de los campesinos cuando van a sus consultas. No tienen que pagarles con estos presentes, pero lo hacen por tradición.

4.14 La yuca y el mundo de las cuevas

La historia oral contiene narraciones sobre la época en que el espiritismo era perseguido y reprimido. En pueblos como Manatí, Yauco, Vega Baja, Ciales, Florida, y otros, he escuchado historias sobre gente que tenía que realizar sus rituales en las cuevas porque "estaba prohibido por el Gobierno".

Muchas cuevas de Puerto Rico se conocían con nombres como "La Capilla", "La Iglesia", "El Convento"; y muchas todavía conservan esos nombres. Son nombres asociados a prácticas religiosas de antaño. Eran prácticas religiosas clandestinas. A nadie se le ocurre pensar que esos nombres surgieron porque el cura del pueblo iba a esos lugares; monte adentro, en zonas aisladas a ofrecer misas. Pero también he conocido personas, espiritistas, identificados como de raza india, que dicen que ellos continuaron haciendo rituales hasta la década de 1940 por tradición. La práctica de llevar casabe a la casa del *bohique* fue suprimida por la Inquisición, pero por dondequiera se ven reductos de esas y otras prácticas suprimidas. Otra práctica que sobrevivió fue la de llevar cemíes a los montes, a las cuevas, a los ríos y quebradas.

4.15 La yuca y los astros

Este tema siempre aflora en las conversaciones con los agricultores de avanzada edad. En los campos de Puerto Rico los ancianos mantenían la tradición de sembrar plantas, arboles, raíces y cualquier otro producto de acuerdo a las fases o ciclos de la luna. Lo mismo ocurría con las cosechas. Agricultores de la zona de Comerío sembraban el "*paslotito*" [*cangre*] de la yuca tan pronto terminaba la luna nueva. A partir de ese momento era que se podía sembrar, pero antes que terminara la fase de cuarto menguante. En el barrio Palomas de Comerío se daban unas yucas tan grandes que... se [referían] a ellas como "*yuconas*". Estas eran de cuatro y cinco libras de peso. Pero si no se siembran en esa fecha, entienden ellos, la mata no pare esa yuconas y son de mala calidad.

En la zona occidental del país, en las montañas del interior, existe una tradición oral que destaca que los tubérculos (todas las plantas cuyos frutos o cultivos crecen debajo de la tierra) deben ser sembrados en luna creciente. Esta tradición fue recopilada por [un] investigador folklorista. . . en la zona de Jayuya y Utuado. La información es muy valiosa pues proviene de una tradición perteneciente a una de las zonas donde la supervivencia indígena es más notable, más evidente; y más corroborable. La yuca, por razones obvias, está directamente relacionada con esta técnica agrícola, que con toda probabilidad se remonta a la sociedad indígena. . . . La litera-tura que trata sobre los conocimientos que tenían los indígenas respecto a la posición de los astros y su relación con el mundo terrestre ha ido en aumento. En este plano o dimensión también podemos ver la relación de la yuca con su cosmovisión espiritual.

4.16 El cangre de la yuca y su dimensión espiritual

Uno de los aspectos de la yuca que más me ha impresionado es el respeto y admiración que los agricultores y sus familias le tenían al "*cangre*". Ya vimos el tema del *cangre* relacionado a las técnicas de sembrar la yuca y su relación con el mundo espiritual. Siempre me llamó la atención que cuando los campesinos mencionaban la palabra su tono de voz cambiaba para expresar cierta veneración. El *cangre* era lo más sagrado de la yuca. Todavía en estos tiempos uno puede escuchar expresiones como esta: "¡Ah, el *cangre*. Eso era lo más grande. Sin el *cangre* no había yuca!" Esas expresiones todavía reflejan el nivel sacro del *cangre*.

El "*cangre*" es el pie de yuca, que sirve de semilla para reproducir la *yucubía*. Esta voz la registró Don Augusto Malaret en su *Diccionario de Americanismos*, pero la registró como una voz cubana. En Puerto Rico se conocía esa voz, pero mayormente la usan los ancianos de la región occidental y sobre todo de la región montañosa. Lo mismo la escucha uno en Florida que en Jayuya. Pero en Comerío, los agricultores que yo he entrevistado no la conocen. Algunos le dicen "*paslote*". . . . Lo interesante es que esa voz, con igual significado se conoce en Colombia y también con la variante de "*cangle*". En el barrio Río Grande de Jayuya un agricultor sembraba los *cangres* en montículos de tierra y basura; sobre todo de hojas, que le servían de abono. El montículo era parecido a los montones que se preparaban desde antes de 1492. Muchos campesinos se presignaban o invocaban algún santo, o decían: "en el nombre del Padre, y del Hijo, y del Espíritu Santo", al sembrarlo.

4.17 El valor sagrado del casabe

Los historiadores, antropólogos, y folkloristas puertorriqueños han realizado una labor encomiable de rescate de información sobre diversos aspectos de nuestra cultura. Cuando se enfrentan a esa extraordinaria cantidad de información concluyen que es el legado de las tres "razas": indios, africanos, y españoles. Sin embargo, la obra que no ha sido concluida es la de adjudicar cuáles elementos pertenecen a tal o cual cultura.

Las madres campesinas. . .tenían la costumbre de mascar los alimentos para que sus infantes pudiesen comerlos y digerirlos. De hecho, la yuca hervida era uno de los alimentos que mascaban para darle a sus bebés. . . .Para mi sorpresa, el casabe era uno de los alimentos mascados por las madres. . . . Esto quiere decir que quinientos años después de la invasión española las jíbaras continuaban alimentando a sus hijos con una dieta totalmente indígena. De la tradición oral se desprende que el casabe no podía botarse, porque el pan era sagrado.

Sabemos por los cronistas que había un casabe conocido con el nombre de "*jaujau*", que era un "pan de casabe de flor, muy blanco y tierno". Los cronistas también lo llamaron "*jabjao*" y a veces aparece escrito "*xabxao*". Aunque en Puerto Rico no logró sobrevivir dicha voz, todavía "se conserva en Venezuela, con igual sentido"[13]. Ese tipo de casabe todavía se producía en Puerto Rico en 1644. Ese año, el Obispo Damián López de Haro, en una carta escrita por él, al referirse al casabe que le sirven destaca lo siguiente: "...y es el pan de esta tierra que la necesidad les ha enseñado á comerlo, pero a mí no me entra de los dientes adentro, aunque lo hacen de diferentes modos y ponen a la mesa uno que es el más florido *jaujao*"[14].

El texto... demuestra que... todavía se producía el *jaujao*. Y también demuestra que el *jaujao* era destinado a la gente de importancia en la sociedad; sobre todo a las figuras representativas del mundo religioso. Si el *jaujao* era destinado al cacique, al *bohique* y a los *nitaínos*, ciento cincuenta años después era elaborado para el Gobernador, el Obispo, los curas y los grandes señores de la aristocracia colonial.

5 Conclusión

5.0 Y mientras continúa el *burén* encendido, recordamos las nobles palabras de nuestra María Teresa Babín:

> ¿Que tal te sientes ahora, don Casabe, entre tantos otros panes exóticos rebanados en serie geométrica, metidos higiénicamente en papel encerado? ¿Te engullían a ti los niños y los hombres de antes con la misma gana que se comen de prisa los de hoy a los otros panes? ¿No te aflige verte relegado al folklore en tu tosca y honrada persistencia mientras reinan en la mesa esos amasijos de blandura esponjosa?
>
> Don Casabe no habla, pero mira, haciendo su presencia esporádica en nuestro apetito. Lo pregonan en San Juan los revendones de Loíza, surge de improviso envuelto en el tumulto oliente de gasolina, y sigue su rumbo sin perder ni la forma ni el gusto. ¿Quién lo prepara todavía? ¿De dónde sale? ¿En que misterioso fogón se le adereza? Mi lengua ávida lo saborea con asombro, como si la magia de un dios tutelar me trasladara al Coayuco y me sintiera recién nacida el año 1511.[15]

PLATE 4
06.16.2002, ISABELA
From CAZABI: *Gift of the Americas*, Volume IV of *12/12*

Manihot Esculenta: Historic and Economic Context

Manihot esculenta: Su contexto histórico y económico

Trish O'Kane

Trish O'Kane is an investigative journalist and a research consultant for non-profit organizations. She has published numerous articles in Spanish and English on politics, culture, and economic issues in the United States and in Latin American media, and is the author of *Guatemala: A Guide to the People, Politics and Culture* (New York: Interlink Books, and London: Latin American Bureau, 1998).

Manihot Esculenta: Historic and Economic Context

In 1492, when Christopher Columbus arrived on Hispaniola, the island that now includes the Republic of Haiti and the Dominican Republic, he found something even more valuable than gold:[1] a bread that allowed European explorers to venture much farther than they had before and to maintain their first settlements in the tropics. While the fifteenth-century sailors had made significant maritime technological advances that enabled them to reach continents on the other side of the Earth, they did not have a foodstuff that would keep well during their long sea voyages. In the tropical heat and humidity of the Caribbean, the Spanish *bizcocho*, biscuits made from wheat, quickly turned into "masses of pulsating weevils."[2] The Taínos, the first people Columbus encountered, introduced him to the technology to produce bread from *Manihot esculenta*, the starch-filled roots of a plant hitherto unknown in Europe.[3]

Commonly referred to as "cassava" or "tapioca" in English, "manioc" in English, French, and Portuguese, and "yuca" in Spanish, *Manihot esculenta* was the staple of the Taíno people. (This food is not to be confused with *Yucca schidigera*, commonly known as yucca, a cactus plant grown in the southwestern United States.) "Cassava" is used throughout this essay, but where the terms "manioc" or "yuca" appear in sources interviewed or cited, the usage is respected.

There are two main varieties of cassava, one sweet and one bitter. The sweet variety deteriorates quickly, but it can be eaten as a vegetable. The Taínos cultivated the bitter variety extensively and, as it is poisonous, they had to subject it to an elaborate process to extract its hydrocyanic acid.[4] The Taínos used the processed bitter cassava to produce a flour from which a variety of foodstuffs could be made. The flour was used to bake a flat bread that the Spanish explorers called "*cazabi*," now known as "*casabe*" or "*cazabe*" in contemporary Spanish, and "cassava bread" or "bammy" in the English-speaking West Indies. Once processed into bread, it could be eaten for as much as a year later if kept dry.

As it did not require packaging or preservatives, "*casabe*" quickly became a desirable commodity among the Spaniards, and galleons stopped off at *casabe* depots where crews filled the holds with it.[5] Columbus had to send his men to stock up on cassava and *casabe* before returning home after his first voyage, as they had little food left.[6] In order to avoid enslavement in gold mines and war with the Spaniards, the Taínos were forced to cultivate cassava and make *casabe* as tribute to the Spanish crown.[7]

More than five centuries later, the root crop is grown in at least ninety-two countries and feeds over half a billion people, according to the Food and Agriculture Organization (FAO), the lead agency of the United Nations for agriculture, forestry, fisheries, and rural development.[8] It is just as valuable to scientists and genetic engineers in San Diego, Ibadan, Dong Nai, Bangkok, and São Paulo as it was to the European explorers: It produces more carbohydrates than any other staple crop, and ranks fourth as a food crop in developing countries after rice, maize, and wheat.[9] At a time when hunger continues to be the world's most serious health problem, and weather changes due to global warming damage the food crops of people in developing countries,[10] cassava is still gold to the intellectual explorers of our age. A gift from the Americas to the world, cassava's geographical and cultural journey indicates how much there is to learn about the natural history of the tropical and subtropical Americas, and the living legacy of the peoples who have lived in the region for thousands of years.

Recent archaeological finds and ethnobotanical discoveries document the early use and cultivation of cassava. In 1997, in a tropical rainforest on a Pacific coastal plain of central Panama, ancient milling stones were found by a team of archaeologists from the University of Pittsburgh. The milling stones are called "edge-ground cobbles" because the grinding facet is on the narrow edge of the stone. Under a stereoscopic microscope, some of the unwashed stones tell an ancient story.[11]

"We recovered thirteen bell-shaped starch grains with combinations of attributes that only occur in manioc roots from an edge-ground cobble," writes archaeologist Dolores Piperno. Piperno and her colleagues' findings of microscopic cassava starch at a settlement called the "Aguadulce Shelter" reveal that between 5,000 and 7,000 years ago, the autochthonous peoples of Panama cultivated many plants, including varieties of cassava that had already been domesticated. The Aguadulce starch grains are the earliest recorded evidence of the root crop in the Americas.[12] Piperno explains: "In the [tropical and subtropical] Americas, more plants were domesticated for their starch-rich underground organs than in any other region of the world. We have long suspected that the lowland tropical forest was one of the ancient centers for the development of prehistoric agriculture."[13]

The finding is significant because it means that the root crop's domestication actually took place much earlier than even the Aguadulce grains show. The plant is not native to Panama and had to have been brought there after domestication via the migration of autochthonous peoples from other regions. Because Brazil is the place where the largest number of species survive today, many scientists believe that the plant originated there and may have been domesticated as long as 9,000 to 10,000 years ago.[14]

Little is known about the earliest stage of agriculture in the Americas, when Brazil's autochthonous peoples must have experimented with and domesticated cassava. They discovered that the bitter variety was poisonous and could be lethal. These ancient cultivators developed a complex process to extract the hydrocyanic acid. Once the cassava had been grated and the resulting pulp squeezed, it could be spread out to dry into flour for cooking or baking. Hunters in Brazil dipped the tips of blow darts in the extracted poison to kill animals quickly, preventing injured animals from escaping and dying elsewhere.[15]

The mythology and cosmology of many of the indigenous peoples who experimented with cassava, and who still use it today, is evidence of its central cultural role. In 1995, Brazilian filmmaker Virginia Valadão immortalized one version of the legend of the first cassava root in the Brazilian documentary *Yákwa, The Banquet of Spirits*.[16] According to the legend, a mother with no food was forced to watch her starving little girl die. The child, whom she buried under the floor of her home, was transformed into the first cassava root. Ethnobotanist Mark J. Plotkin writes that this is where the word "manioc" originated, coming from "*mani oca,*" which means "wood spirit root."[17]

For the Tukanoans of the northwest Amazon, the root is central to their creation story. Plotkin says the Tukanoans believe that the first man and woman came to Earth from the Milky Way, traveling in a canoe drawn by a sacred anaconda: "In the canoe they carried the three plants necessary for life in the rain forest: *yagé*, a hallucinogenic vine that allowed them to communicate with the spirit world; *coca*, a stimulant that helped them work and hunt without fatigue; and cassava, their staple foodstuff."[18]

Casabe in the Greater Antilles: Sixteenth-Century Bread Depots and the Bread Tax

Nine millennia after it was first domesticated, cassava entered the known written record with the arrival of Europeans financed by the Spanish Crown. Columbus established his headquarters on the Greater Antillean island of Hispaniola in 1492, on the site that is now the city of Santo Domingo. He and his men were very impressed with Taíno agriculture, and in the log of his first journey, he compared Hispaniola's fields to Spain's: "In all Castile there is no land that can be compared to this in beauty and fertility. All this island and the Isla de la Tortuga are entirely cultivated like the plain of Córdoba."[19]

Columbus, who made the first European documentation of the cultivation of cassava, wrote: "The Indians sow little shoots, from which small roots grow that look like carrots. They serve this as bread, by grating and kneading it, then baking it in the fire. They plant a small shoot from the same root again in another place, and once more, it produces four or five of these roots. They are very palatable and taste exactly like chestnuts. The ones grown here are the largest and best I have seen anywhere."[20]

Since the food the Spanish brought with them spoiled in the tropical heat, and many of the Old World plants did not adapt to the tropics, food became a contentious issue between the explorers and the Taínos. Taíno rulers such as Guarionex complained that a Spaniard seemed to consume much more than a Taíno, and that the newcomers did not understand the harvesting cycles of their root crop. They not only ate the food ready to be harvested, they also consumed manioc that should have stayed in the ground for another six months. Famine followed, and conflict simmered.[21]

The Taínos immortalized the central role of *casabe* in their dealings with the Spanish in a series of twelve hundred black-charcoal pictographs on the walls of a limestone cave in the eastern region of the Dominican Republic.[22] Archaeologists John W. Foster and Adolfo López Belando describe the paintings as "a significant storehouse of ancient Taíno knowledge—a university in picture writing."[23]

One panel twelve meters long and one meter high begins with the image of a grater used to grind cassava to make *casabe*.[24] It is followed by a pictograph of a barbecue where the bread dough was baked, as a Taíno leader in a headdress looks on. The next scene shows Taínos loading the *casabe* onto a Spanish longboat to pay taxes to feed the Spaniards.[25] López Belando writes: "This assured their independence in the face of a voracious Spanish administration eagerly seeking slaves to work in the gold mines."[26] A longboat is shown being loaded and then travelling back to Spain.[27]

The tribute scene takes place on the small island of Saona, a major bread depot that supplied the first Spanish settlements.[28] Saona is in the Higüey region of the Dominican Republic, at that time, the easternmost chiefdom on the island of Hispaniola, and the last area to succumb to Spanish rule in 1504. *Casabe* played a central role in the fall of Higüey in 1502, when a Spanish attack dog killed a Taíno leader while the bread was loaded onto a Spanish galleon.[29] The killing provoked another war between the Spaniards and Taínos in the Higüey region. Historian Samuel Turner writes that the Taínos of Higüey sent peace envoys to the Spanish, offering to serve them if the Spanish stopped hunting them: "The newly subjugated Indians were commanded to lay out a vast field of yuca from which they would make *cazabe* bread for the King and, upon executing this command, were to be free to live in their land and not be obliged to come and serve in Santo Domingo. It is probable the Spaniards were as eager to end the war as the Indians. Santo Domingo, from its founding, was largely dependent on bread imports from both Saona island and greater Higüey. The war left bread production and procurement mechanisms in complete disarray and with the large influx of Spaniards...only months before, Santo Domingo went through a period of severe food shortage."[30]

Peace did not last however, and *casabe* once again fueled the conflict. In 1504, the Taínos were forced to transport *casabe* to the first Spanish city in the New World, Santo Domingo.[31] This, in addition to other Spanish abuses, provoked a Taíno attack on a Spanish fort. The second and final war of Higüey began. Many Spaniards starved to death on the island during this war because they did not cultivate any food themselves on the island, and had become dependent on Taíno food production and expensive imports from Spain.[32]

When the Spanish finally captured and hanged Higüey leader Cotubanamá in 1504, they believed they controlled Hispaniola. However, many Taínos had already escaped and were living in areas of the island that the Spanish could not police. Anthropologist Lynne Guitar, a specialist on the Taínos, explains: "What's left out of all the history is indigenous peoples' efforts to maintain their autonomy. The Taínos developed an economic relationship with the Spaniards. They were not totally devastated by the arrival of Europeans. They did everything to maintain their own identity and economy. Eventually they became Dominicans, and Dominicans are not Spaniards."[33] For both rich and poor, the staple of the autochthonous peoples of the Greater Antilles of the fifteenth century is an integral part of the modern Dominican diet, sold in most grocery stores and served with meals in the Presidential Palace.[34]

Feeding Africa and Asia

While it was the Taínos who showed the Spanish the value of cassava, it was the Tupinambá people of eastern Brazil who taught the Portuguese how to process and utilize it for baking.[35] According to the FAO, Nigeria has now surpassed Brazil and is the largest producer of the root crop in the world today. The fact that an African country leads the modern world in production dates back to the first encounters between Portuguese explorers and the Tupinambá in the sixteenth century, and then the subsequent journeys of Portuguese ships to Africa, when the Portuguese took *farinha*, or cassava flour, with them as provisions.[36] The Portuguese, Spanish,

French, Dutch, and English all spread cultivation to different areas of tropical and subtropical Asia and Africa over the next three centuries.

Various European explorers mention cassava in their chronicles of Asian and African exploration. The Dutch explorer Olfert Dapper, for example, describes Luanda, Angola, in 1640 as a primary production center for cassava. Dapper wrote that the Portuguese pressured Africans to produce the crop to insure a food supply for the town, and encouraged cultivation because it thrived even in infertile soil.[37]

Documentation regarding the precise dates that cultivation spread across Africa is scarce, but we do know how it spread. The International Institute of Tropical Agriculture (IITA) in sub-Saharan Africa attributes the successful cultivation of cassava to several mechanisms such as Portuguese-Brazilian contact, river travel in Africa, overland trade, and mass migration. The journey of cassava from the Americas to Africa and the East between the sixteenth and nineteenth centuries was mainly over water, and river trade was of particular importance: "Cassava was especially suited to take along on trips, presumably in processed form such as '*chikwangue*,' and constituted a balanced diet in combination with fish."[38]

Mary Karasch, an expert on Brazil and author of the entry on manioc in *The Cambridge World History of Food*, insists that freed slaves who returned to Africa from the Americas were instrumental in the dissemination of the bitter variety of the root in Africa in the early nineteenth century. The IITA also mentions the "catalytic effect" of freed Brazilian slaves who returned to West Africa around 1800, promoting production in Nigeria and Benin.[39] Colonial administrators encouraged cultivation of the root in the nineteenth and twentieth centuries because of its value as a famine reserve crop and its resistance to locust plagues and drought. Cultivation also spread during the twentieth century because cassava can be grown under a wide variety of circumstances.

Four centuries after its introduction in Africa, cassava is still mainly a foodstuff. In West Africa, it is eaten in many forms, including as *foufou*, a cassava-based paste, and as a dried convenience food called *gari*. Cassava leaves are also a valued ingredient in African cuisine; these high-protein greens are used to thicken stews and sauces.[40]

In Asia, utilization evolved differently, and today the crop is a vital raw material for various industries, in addition to serving as a foodstuff. The root crop entered Asia through the islands in the western Indian Ocean, which were central ports of commerce by the ninth century for Indonesian, Indian, and Arab traders.

Karasch describes the island of Mauritius in the Indian Ocean as a distribution point for manioc into Sri Lanka (formerly Ceylon), where the root crop was introduced in 1786. The French first claimed the island of Mauritius as "Ile de France" in 1715, and around 1736, French governor Mahé de la Bourdonnais had manioc brought from Brazil to guarantee a reliable food source.[41] The French also introduced it to the French islands near Madagascar, and later, the technology was transferred to Madagascar, where it became a major staple.[42]

A Dutch gardener named Johannes Elias Teysmann helped introduce the plant to Indonesia. In 1830, Teysmann became the curator of the Bogor Botanic Garden in Bogor, Indonesia, a major center for the promotion of agriculture and horticulture. When the plant was found growing as a hedge on Batam, an island off Sumatra, Teysmann promoted its cultivation in Indonesia for use as an alternative food source when rice harvests failed.[43]

It was the Spanish who took cassava to the Philippines, where it is called *balingboy*, *Kamoteng-Kahoy*, or *Kamoteng-moro*. By the year 1800, Karasch writes, manioc was being grown from Ceylon to the Philippines. Although it never replaced rice as the staple crop of Asia, it became very popular in countries and regions where good farmland was scarce.

Cassava Today

Five hundred years after the Taínos introduced Columbus to *casabe*, the crop yield of Asia has far surpassed yields in the Americas. In 1997, according to the International Center for Tropical Agriculture (CIAT), headquartered in Colombia, the root crop covered 3.9 million hectares in Asia, more than in Latin America. Because of the plant's capacity to grow long feeder roots (distinct from the tuberous roots) that can find nourishment 50–100 centimeters below the surface,[44] it is planted mostly by Southeast Asia's poorest farmers who live on marginal lands.[45] Asia's major producers are Thailand, Indonesia, India, southern China, and Vietnam, respectively.

Thailand currently leads the world in cassava starch production.[46] In the 1970s, this country began exporting massive quantities of dried cassava chips and pellets to the Europe for use in animal feed. Cassava starch is an important ingredient in processed foods, paper, textiles, and even pharmaceuticals. Indonesia and Vietnam are following the Thai model of increasing cultivation and production. In Hanoi, cassava is processed in large, modern factories. Once considered a crop of "last resort" for consumption during wars and famines, the starch is now used predominantly to produce monosodium glutamate in Vietnam. Currently, more than half of the land used to grow the crop is in Africa.[47] FAO forecasts that global production of cassava will rise to nearly 210 million tons per year by 2005.[48]

However, cassava's contemporary story worldwide is not a simple success story. As with other staple crops, its performance in a country seems to depend on that nation's infrastructure and economic strength, and the ability of a farmer to buy certain inputs. Problems with cultivation and overproduction in Africa and Asia must be considered before it is promoted as a solution for endemic nutritional problems. According to soil experts, there is an ecological downside to the Asian cassava boom, since most of the farmers planting the root are poor and live on fragile uplands. Increased cultivation has also increased the risk of further soil erosion.[49]

In Africa, the crop has also been attacked in recent years by a series of new pests, causing famine in some countries.[50] Another complication on that continent is recurrent outbreaks of Konzo disease. This motor neuron disease, which can cause paralysis and blindness, primarily attacks children and women of childbearing age, and the very poor and undernourished, particularly those with an insufficient sulfur intake. Named for the area in southern Democratic Republic of the Congo (formerly Zaire), where this illness first appeared in the 1930s, Konzo strikes when a starving population relies solely on the bitter variety of the root as their staple and the crop has not been processed properly.[51] Outbreaks have occurred in rural Africa, particularly the Central African Republic, Mozambique, Tanzania, and the Democratic Republic of the Congo.[52]

While cassava was a staple crop for the Taínos, Columbus and his fellow explorers noted that they also ate a rich and varied diet, accompanying their *casabe* with many types of fresh seafood and other protein. The Taíno example provides a valuable lesson: Reliance on a monoculture has proven time and again to be problematic, as this type of agriculture increases nutritional, economic, and environmental vulnerability.[53]

One of the most eminent Spanish chroniclers of the sixteenth century, Gonzalo Fernández de Oviedo y Valdés, wrote about "the seven noteworthy characteristics of cassava." In 2003, the food with many names also had over 300 industrial uses.

In Brazil, its probable place of origin, it is consumed in all regions, by all classes. The middle class enjoys roasted manioc with steak and French fries. The poor boil it in fish stews or roast it. The Casa do Pão de Queijo, a popular Brazilian fast-food chain with 141 franchises, serves cassava cheese bread.[54] The "gift of the Americas" is eaten for breakfast in India as *puttu*, a traditional steamed dish, and is served in *bibingka* pancakes in immigrant Filipino communities in New York. Eel farmers in Thailand depend on the transparent gelatinous starch as a cold, water-soluble binder, and all over Asia, the starch sweetens and thickens desserts. How many North Americans have enjoyed tapioca pudding, and realized that its journey to our table started over nine thousand years ago in Brazil?

Manihot esculenta: Su contexto histórico y económico

En 1492, cuando Cristóbal Colón llegó a Hispaniola, la isla que hoy se divide entre la República de Haití y la República Dominicana, encontró algo más valioso incluso que el oro[1]: un alimento que permitió que los exploradores europeos emprendieran viajes mucho más largos que antes, y que sostenía a los primeros poblados españoles en el trópico. Gracias a los avances tecnológicos marítimos de la época, los marineros del siglo XV lograron alcanzar tierras desconocidas al otro extremo del planeta. Sin embargo, no poseían un comestible que pudiera sobrevivir bien los largos viajes en el mar, más el calor húmedo del trópico Caribe. En el nuevo clima, el bizcocho español, una especie de galleta de trigo, se deshacía rápidamente y se convertía en "una masa de gorgojos pulsantes"[2]. Fueron los taínos, el primer pueblo que Colón conoció en el nuevo mundo, quienes enseñaron a Colón la tecnología para producir un pan de *Manihot esculenta*, una planta desconocida en Europa en aquel entonces, con raíces de almidón[3].

Conocido como "*cassava*" o "*tapioca*" en inglés, "*manioc*" en inglés, francés, y portugués, y "yuca" en español, *Manihot esculenta* era el verdadero pan del pueblo taíno. (No se debe confundir este alimento con *Yucca schidigera*, conocido como *yucca*, una planta cultivada en el suroeste de Estados Unidos). Usamos "yuca" en este ensayo pero se ha respetado el uso de los términos "*manioc*" o "*cassava*" según su utilización en fuentes originales, entrevistas o documentos.

Existen dos variedades principales de yuca: la dulce y la amarga. La variedad dulce se deteriora rápidamente, aunque se puede hervir y comer como vegetal. Los taínos cultivaban la variedad amarga en grandes extensiones y, como es venenosa, la sometían a un proceso complicado para sacar el ácido hidrociánico[4]. De esta pulpa procesada, los taínos producían una harina, y de esta harina varios tipos de comida. Los españoles llamaban "cazabi" al pan que los taínos hacían con la harina, conocido ahora como "casabe" o "cazabe" en el mundo hispanohablante contemporáneo, y como "*cassava bread*" o "*bammy*" en las islas caribeñas donde se habla inglés. El pan duraba hasta un año si se mantenía seco.

Entre los españoles, este pan milagroso que no requería medidas de conservación especial, pronto se convirtió en una provisión indispensable. Las carabelas empezaban a parar con frecuencia en los centros de producción de casabe para que sus tripulaciones pudieran llenar sus bodegas[5]. Antes de volver a casa después del primer viaje al nuevo mundo, Colón tuvo que mandar a sus propios marineros en búsqueda de la yuca y el casabe porque ya les faltaba comida[6]. Para evitar una guerra con los españoles y la esclavitud en las minas de oro, los taínos fueron forzados a cultivar la yuca y producir casabe para poder pagar los impuestos[7].

Cinco siglos más tarde, los agricultores de noventa y dos países siembran la raíz, alimentando a más de 500 millones de personas, según la Organización para la Agricultura y la Alimentación de las Naciones Unidas (FAO)[8]. Los científicos e ingenieros genéticos de San Diego, Ibadan, Dong Nai, Bangkok, y São Paulo hoy la valoran tanto como los exploradores europeos de hace 500 años, porque produce más carbohidratos que cualquier otra cosecha. Para los países en vías de desarrollo, es el cuarto cultivo más importante, después del arroz, el maíz, y el trigo[9]. En esta época moderna, cuando el hambre sigue siendo el problema de salud más grave del mundo, y los cambios climáticos causados por el efecto invernadero dañan a los cultivos de muchos países pobres[10], la yuca es el oro de los exploradores intelectuales de nuestra época. Un obsequio de las Américas al mundo, el viaje geográfico y cultural de la yuca nos enseña cuán poco sabemos de la historia natural de la América tropical y subtropical, y del legado dinámico y persistente de los pueblos autóctonos que han residido en esta región por miles de años.

Recientemente, algunos arqueólogos y etnobotánicos han descubierto nuevas pistas que dan luces sobre el uso y cultivo de la yuca en la antigua América. En 1997, en un bosque tropical en el centro de Panamá, un equipo de arqueólogos de la Universidad de Pittsburg encontró piedras viejísimas utilizadas para procesar la raíz. Bajo un microscopio, algunas de estas piedras cuentan una historia milenaria[11].

"Encontramos trece granos de almidón con ciertas combinaciones de atributos que sólo ocurren en las raíces de *manioc*", escribe la arqueóloga Dolores Piperno. Los granos microscópicos del almidón de la yuca en el poblado de "Aguadulce Shelter" revelan que hace 5.000 a 7.000 años, los pueblos autóctonos de Panamá tenían muchos cultivos, incluyendo variedades de la raíz ya domesticadas. Los granos de Aguadulce son la prueba más antigua de la cultivación de la yuca en las Américas[12]. Piperno explica: "En las Américas [tropical y subtropical], se domesticaban más plantas por sus órganos subterráneos con un contenido alto en almidón, que en cualquier otra región del mundo. Por mucho tiempo hemos sospechado que el bosque tropical bajo era uno de los centros de desarrollo de la agricultura prehistórica"[13].

El hallazgo de Aguadulce es importante porque significa que alguien domesticó esta planta mucho antes incluso que la época de Aguadulce. La raíz no se originó en Panamá y fue traída por algún viajero a través de la migración de los pueblos autóctonos de otras regiones. Hoy en día, Brasil es el lugar donde sobreviven el mayor número de especies y, por lo tanto, muchos científicos creen que es su punto de origen. También opinan que la planta pudo haber sido domesticada hace 9.000 o 10.000 años[14].

Existe poca información sobre esta primera etapa de la agricultura en nuestro continente, cuando los pueblos autóctonos de Brasil experimentaban con la planta y la hicieron comestible. Alguien persistente y emprendedor descubrió que la variedad amarga era venenosa y su ingestión podría ser hasta mortal. Estos primeros agricultores desarrollaron un proceso complejo de mojar a la raíz, lavarla, y rallarla hasta extraer su veneno. Después de sacar el jugo de la pulpa, se dejaba secar y luego se hacía una harina para cocinar y hornear. Nada de desperdicios. Hasta el veneno servía a los antiguos cazadores de Brasil, quienes mojaban la punta de sus dardos en el líquido para asegurarse de que los animales murieran más rápido, así evitando la huída de los heridos[15].

Los pueblos indígenas que experimentaban con la yuca, y que siguen consumiéndola, comparten una cosmovisión y mitología, donde la raíz juega un papel central. En 1995, la cineasta brasileña Virginia Valadão inmortalizó una versión de la leyenda de la primera raíz de cassava en el documental brasileño *Yákwa, The Banquet of Spirits*[16]. Según esta leyenda, una madre hambrienta tuvo que ver a su hija pequeña morir de hambre. Ella enterró a su hija debajo del piso de la casa, y luego la niña se transformó en la primera raíz de yuca. El etnobotánico Mark J. Plotkin escribe que la palabra "*manioc*" surge de esta leyenda, porque en el idioma indígena "*mani oca*" quiere decir "raíz del espíritu de madera"[17].

Para los Tukanoans del noreste de Amazonas, la raíz es el centro de su historia de creación. Según Plotkin, los Tukanoans creen que el primer hombre y la primera mujer vinieron a la Tierra desde la Vía Láctea, en una canoa arrastrada por una anaconda sagrada: "En la canoa llevaban las tres plantas esenciales para sobrevivir en la selva: *yagé*, un alucinógeno para comunicarse con el mundo de los espíritus; la coca, un estimulante para cazar sin fatigarse; y la yuca, su alimento principal"[18].

Casabe en las Antillas Mayores: Las panaderías del siglo XVI y el pan como tributo

Nueve milenios después de ser domesticada, la yuca entra a la historia escrita con el arribo de los europeos financiados por la corona Española. Colón estableció su sede en la isla antillana mayor de Hispaniola en 1492, en el lugar hoy llamado Santo Domingo. Se quedó muy impresionado por la agricultura taína, y en su diario del primer viaje compara los campos cultivados de Hispaniola a los de España, en cuanto a su belleza y fertilidad.

Los alimentos no tardaron en convertirse en tema conflictivo entre exploradores y anfitriones taínos. La comida española no resistía el clima tropical y muchos cultivos del viejo mundo tampoco prosperaban en el trópico. Algunos *caciques* taínos se quejaban que un español comía mucho más que un taíno, y que los huéspedes europeos no entendían los ciclos de la cosecha. Los españoles no sólo comían las raíces cosechadas, sino también consumían las que aún necesitaban seis meses más para madurar bajo la tierra, según una queja específica del *cacique* Guarionex. La hambruna llegaba, y los conflictos crecían[19].

Los taínos inmortalizaron el papel clave que el casabe jugaba en su relación con los españoles en una serie de 1.200 representaciones pictográficas en una cueva totalmente oscura en el este de la República Dominicana[20]. Los arqueólogos John W. Foster y Adolfo López Belando describen las pinturas de carbón en la Cueva de José María como "un almacén importante de la sabiduría taína antigua—una universidad en representaciones pictográficas"[21].

Una sección de doce metros de largo y un metro de alto empieza con la imagen de un rallador que se utilizaba para convertir la yuca en casabe[22]. En otra escena, hay una barbacoa donde se hornea la masa, mientras un *cacique* taíno está de guardia. La próxima escena muestra a los taínos cargando el casabe y llevándolo a una carabela española para pagar los impuestos que eran el sustento de los españoles[23]. El arqueólogo López Belando escribe: "Esto les aseguraba su independencia frente a la voraz administración española siempre ávida de esclavos para el trabajo en las minas de oro"[24]. Otra escena enseña a un barco español que espera su carga de pan para emprender su viaje a España[25].

La escena retrata a una transacción en la pequeña isla de Saona, un centro de producción de pan que abastecía a los primeros pueblos españoles[26]. Saona está en la región de Higüey de la

República Dominicana: el antiguo cacicazgo más al este de la isla de Hispaniola y el último que cayó en manos de los españoles en 1504. El casabe jugó un papel central y trágico en cuanto a la caída de Higüey, empezando en 1502 cuando un perro español de ataque mató a un *cacique* taíno mientras los taínos llevaban el pan a una carabela española[27]. La matanza provocó otra guerra entre españoles y taínos en la región de Higüey. El historiador Samuel Turner escribe que los taínos de la región mandaron sus enviados de paz a los españoles, con el mensaje de que los taínos les pagarían el tributo si los españoles dejaban de cazarles como animales: "Los indios recién subyugados recibieron la orden de cultivar un campo enorme de yuca, de lo cual harían el casabe para el Rey, y después de ejecutar esta orden, serían libres y podrían vivir en sus tierras sin la obligación de servir (a los españoles) en Santo Domingo. Es probable que los españoles anhelaran terminar con la guerra tanto como los indios. Desde que se estableció, la ciudad de Santo Domingo dependía mayormente del pan importado desde la isla de Saona y Higüey. La guerra interrumpía completamente a la producción de pan y los mecanismos de provisión y con el gran influjo de españoles...solamente unos meses antes, Santo Domingo pasó por un período de hambruna severa"[28].

La paz era efímera, y otra vez el casabe calentaba el conflicto. En 1504, los españoles forzaron a los taínos a transportar el casabe a Santo Domingo, la primera ciudad española del "nuevo mundo"[29]. Provocados tanto por esta situación, como por los otros abusos de los españoles, los taínos atacaron a una fortaleza. Así empezó la segunda y última guerra de Higüey. Muchos españoles murieron de hambre en la isla durante esta guerra porque ellos mismos no cultivaban. Para comer dependían demasiado de la producción alimenticia taína y de las importaciones costosas de España[30].

En 1504, cuando los españoles finalmente capturaron y ahorcaron a Cotubanamá, el *cacique* de Higüey, creían que ya controlaban la isla de Hispaniola. Sin embargo, muchos taínos ya habían huido y vivían en zonas de la isla fuera del control de los españoles. La antropóloga Lynne Guitar, especialista en la historia taína explica en una entrevista: "Lo que queda fuera de la historia es el esfuerzo del pueblo indígena para mantener su autonomía. Los taínos desarrollaron una relación económica con los españoles. No quedaron arrasados con la llegada de los europeos. Hicieron todo lo posible para mantener su identidad y su economía. Eventualmente llegaron a ser dominicanos y los dominicanos no son españoles"[31]. Para ricos y pobres, la comida de los pueblos autóctonos de las Antillas Mayores del siglo XV es parte integral de la cocina dominicana moderna. Se vende en tiendas, supermercados y hasta acompaña a platos típicos en el Palacio Nacional[32].

América abastece a África y Asia

Mientras los taínos enseñaron a los españoles el valor de la yuca, el pueblo tupinambá del este de Brasil enseñó a los portugueses como procesarla y utilizarla para hornear[33]. Según la FAO, Nigeria es el productor más grande del mundo hoy en día, sobrepasando a Brasil. El hecho de que un país africano domine la producción moderna se debe a estos primeros encuentros entre exploradores portugueses y los tupinambá en el siglo XVI, y los viajes posteriores, cuando los portugueses volvieron a África y llevaban la *farinha* para comer, una harina hecha de *manioc*[34]. Durante los tres siglos siguientes, los comerciantes y exploradores de Portugal, España, Francia,

Holanda, e Inglaterra propagaron el cultivo de la yuca en diferentes regiones del trópico y subtrópico de Asia y África.

Varios exploradores europeos mencionaron la yuca en sus crónicas de exploración. Olfert Dapper, el explorador holandés, describe a Luanda, Angola, en 1640 como un centro importante de producción de yuca. Dapper escribió que los portugueses presionaron a los africanos a cultivar la raíz para asegurar el abastecimiento del pueblo, y también porque crecía hasta en tierras estériles[35].

No sabemos exactamente cuándo se extendió por África por la falta de documentación histórica, pero si sabemos cómo. El Instituto Internacional de Agricultura Tropical (IITA-International Institute of Tropical Agriculture) en la sub-Sahara, atribuye la exitosa cultivación de la cassava a varios mecanismos como el contacto portugués-brasileño, el tráfico fluvial en África, el comercio y la migración masiva. El recorrido desde las Américas al otro lado del planeta era principalmente por vía acuática, y el tráfico fluvial era de suma importancia: "La cassava era particularmente conveniente para viajes, presumiblemente en formas procesadas como el '*chikwangue*', y constituía una dieta balanceada en combinación con el pescado"[36].

Mary Karasch, una experta sobre Brasil y la autora de la sección "*manioc*" en *The Cambridge World History of Food* insiste en que los esclavos liberados, quienes regresaron a África desde las Américas, jugaron un papel primordial en cuanto a la diseminación de la variedad amarga en el siglo XIX. La IITA también menciona "el efecto catalítico" que tuvieron los esclavos brasileños liberados quienes retornaron al oeste de África alrededor de 1800, promoviendo la producción en Nigeria y Benin[37]. Los administradores coloniales fomentaban la siembra en los siglos XIX y XX por su valor como cultivo de reserva en caso de hambruna, y por su resistencia a plagas y la sequía. El cultivo también extendió durante el siglo XX por la capacidad de la planta de prosperar en diversas circunstancias climáticas y agrícolas.

Cuatro siglos después de su introducción en África, la raíz sigue sirviendo primordialmente para el consumo. En el oeste del continente se come de diversas maneras como en el *foufou*, una masa a base de yuca que es muy popular, y seca en una comida rápida llamada *gari*. Las hojas de la planta tienen un alto contenido de proteínas, y es un ingrediente en la cocina africana para dar sustancia y sabor a los guisados y las salsas[38].

En Asia, el uso de la raíz evolucionó de otra manera. Además del consumo, varias industrias modernas dependen de ella como materia prima. La planta entró a Asia por varias puertas. Las islas del oeste del Océano Indico fueron una de las puertas principales por ser un centro de comercio en el siglo IX para los árabes, indios e indoneses.

Karasch describe la isla de Mauritius en el Océano Indico como un punto de diseminación de manioc, para introducirla en 1786 en Sri Lanka (la antigua Ceylón.) En 1715, los franceses tomaron posesión de la isla, llamándola "Ile de France", y alrededor de 1736, el gobernador francés Mahé de la Bourdonnais, introdujo la *manioc* desde Brasil para garantizar una fuente confiable de comida.[39] Los franceses también la llevaron a las islas francesas cerca de Madagascar, y luego esta tecnología fue transferida a esta última isla donde se convirtió en un cultivo principal[40].

Un jardinero holandés de nombre Johannes Elias Teysmann ayudó a introducir la planta en Indonesia. En 1830, Teysmann fue nombrado el administrador del Jardín Botánico de Bogor, Indonesia, un centro importante de agricultura y horticultura. Cuando la planta fue encontrada creciendo como una cerca en la isla de Batam, Teysmann, preocupado por el abastecimiento cuando fallaba la cosecha de arroz, promovió la *manioc* como cultivo alternativo en Indonesia[41].

Los españoles la trasladaron a Filipinas, donde se llama *balingboy*, *kamoteng-Kahoy* o *kamoteng-moro*. Ya para el año 1800, escribe Karasch, se cultivaba desde Ceylón hasta Filipinas. Aunque nunca reemplazó al arroz como cultivo principal de Asia, muchos agricultores escogieron la *manioc* como una opción en países y regiones donde había poca tierra fértil.

La yuca hoy en día

Quinientos años después de que los taínos enseñaron a Colón el pan de casabe, el rendimiento en Asia era mucho mayor que en las Américas. En 1997, según el Centro Internacional de Agricultura Tropical (CIAT) con sede en Bogotá, Colombia, la raíz cubría 3,9 millones hectáreas en Asia, más que en América Latina. Los campesinos asiáticos más pobres que viven en tierras marginales[42] son quienes más la utilizan por su capacidad de desarrollar raíces (distintas a las raíces tubérculos) muy largas que pueden encontrar su sustento 50 a 100 centímetros debajo de la tierra[43]. Los productores más importantes de Asia son Tailandia, Indonesia, India, el sur de China, y Vietnam, respectivamente.

El líder mundial en la producción de almidón de la raíz es Tailandia[44]. En la década de los 70, este país empezó a exportar cantidades masivas de bolitas y támaras de *cassava* a la Unión Europea para comida de animales. El almidón es un ingrediente clave en la comida procesada, el papel, los textiles y hasta los farmacéuticos. Indonesia y Vietnam han adoptado el modelo Tailandés en cuanto a incrementar el cultivo y la producción. En Hanoi, la raíz es procesada en grandes fábricas modernas. Antes considerada como cultivo de "última instancia" para consumir durante guerras y hambrunas, ahora el almidón se utiliza para producir el monosodium glutamate en Vietnam. Más de la mitad de la tierra que se utiliza para cultivar la raíz está en África[45]. Según la FAO, para el año 2005 la producción mundial de *cassava* será casi 210 millones de toneladas anuales[46].

Sin embargo, la historia contemporánea de esta raíz tan fascinante es compleja, y no se puede caracterizar como un éxito total. Su desempeño, como con otros cultivos, depende en gran parte de la infraestructura del país, la fuerza de su economía y la capacidad de un agricultor de comprar ciertos insumos. Antes de promover este cultivo como la solución a la desnutrición endémica en África y Asia, se deben considerar los problemas de sobre-cultivación y sobre-producción que ya se están dando. Según algunos expertos en suelos, el *boom* de *cassava* en Asia ha tenido consecuencias ecológicas negativas, porque la mayoría de los agricultores que lo siembran son pobres y viven en tierras frágiles. Entre más se cultive, también más riesgo hay de desgastar la tierra[47].

En África, una serie de plagas nuevas han atacado al cultivo, provocando hambruna en varios países[48]. Los recurrentes brotes de la enfermedad de "Konzo" representan todavía otra complicación en este continente. Nombrada por el área en el sur de la República Democrática del Congo donde apareció en los años 30, la enfermedad ataca a una población hambrienta que solamente come la variedad amarga de la yuca, y no la ha procesado bien[49]. La enfermedad causa parálisis y ceguera, atacando a niños y mujeres jóvenes, los muy pobres y mal nutridos, particularmente los que sufren de una deficiencia de sulfuro. Han ocurrido brotes en zonas rurales de África, particularmente en la República Central de África, Mozambique, Tanzania, y la República Democrática del Congo[50].

Mientras la yuca era un cultivo importante de los taínos, Colón y sus compañeros notaron que los indígenas también comían una dieta rica y variada, y acompañaban su casabe con mariscos, pescado, y otras proteínas. El ejemplo taíno nos ofrece una lección: se ha demostrado una y otra vez que la dependencia al monocultivo es problemática, porque este tipo de agricultura incrementa la vulnerabilidad nutricional, económica, y ecológica[51].

Uno de los cronistas españoles más eminentes del siglo XVI, Gonzalo Fernández de Oviedo y Valdés, escribió sobre "siete las características notables que se encuentran en la yuca". En 2003, la raíz de muchos nombres también se utilizaba en más de 300 productos industriales.

En Brasil, su probable lugar de origen, tanto ricos como pobres lo consumen en todas las regiones. La clase media disfruta un plato de *manioc* asada con carne y papas fritas. Los pobres la hierven con pescado en un guiso, y también la comen asada. La Casa do Pão de Queijo, una popular cadena brasileña de comida rápida, con 141 sucursales, vende pan de queso y *cassava*[52]. Este "obsequio de las Américas" se come para el desayuno en la India como *puttu*, un plato tradicional al vapor, y se sirve en las panquecas de *bibingka* en comunidades filipinas en Nueva York. Los criadores de anguilas en Tailandia requieren el almidón gelatinoso y transparente para aglutinar el agua fría de sus estanques, y en toda Asia endulza y da cuerpo a postres. ¿Cuántos norteamericanos han disfrutado el pudín de "*tapioca*" sin darse cuenta de que su largo viaje a la mesa del comedor empezó hace 10.000 años en Brasil?

PLATE 9
06.16.2002, ISABELA
From CAZABI: *Gift of the Americas*, Volume IV of *12/12*

Afterword

For the past ten years, InterAmericas and the Research Institute for the Study of Man (RISM) have developed and co-sponsored exhibitions and publications that reintegrate the arts and humanities with the social sciences. The archives of Phase I of THE NEW OLD WORLD/*El nuevo viejo mundo* have no real precedent in the arts or social sciences and constitute new information from which fresh knowledge can be derived by researchers in various disciplines. RISM, as co-publisher with InterAmericas of Volume IV of *12/12*, CAZABI: *Gift of the Americas*, is primarily interested in understanding the cultural factors surrounding the traditional cultivation and processing of cassava, which evolved to eliminate the potentially toxic concentrations of cyanogenetic glucosides, and efficiently and effectively provide carbohydrates for a balanced diet.

The selections in this publication document the centrality of the techniques for the detoxification of cassava and its proper food combinations to the culture of Amerindians of the insular Caribbean. Increasingly, social scientists recognize that the integration of ritual custom and practice in food preparation and consumption is an efficient cultural mechanism for transmitting essential information necessary for survival. The introduction of cassava into Africa and Asia seems to have been successful where it was compatible with the dietary customs and practices of the indigenous people of the Americas. Where this compatibility does not occur, the effects can be devastating, as evidenced in outbreaks of kwashiorkor (a potentially fatal disease with multiple symptoms caused by protein deficiency) or Konzo (a motor neuron disease) in the Democratic Republic of Congo, where the cassava detoxification process is expedited in times of drought or famine.

The global expansion of the cultivation of cassava without the attendant cultural context has had serious health consequences, including disease and malnutrition, where a diet based on cassava is insufficient in protein and/or iodine, particularly when the root is under-processed. Modern biochemistry and genetics have shown that cyanogenetic glucosides in cassava, along with a naturally occurring enzyme called linamarase, produce the neurotoxin hydrogen cyanide. A review of the medical literature indicates that these glucosides affect the bioavailability of micronutrients, including iodine and the sulfur-containing amino acids methionine and cysteine (which are involved in the processing of iodine). Even though cassava produces one of the highest yields of starch per hectare, it is low in protein, and amino acids are the basic building blocks of protein. A diet in which cassava is eaten alone will result in protein malnutrition and suppression of the thyroid gland's ability to utilize iodine.

Amerindians and their modern descendants consume cassava and/or cassava bread with fish, which is high in both protein and iodine.

The capacity to maintain "sustainable economic development," which is essential to the survival of the burgeoning populations in developing countries, may depend on the appreciation by modern "Western" science of ancient practices, in both the culture of origin and the culture into which the food is being introduced. In this light, many development initiatives by Europeans and North Americans may require reassessment. RISM encourages research in the social sciences and medicine to enable traditional knowledge to be incorporated with the introduction of new food crops for mutual benefit and growth. Revisiting the history of the documentation of "bread made from yuca" may set a precedent for consideration of cultural factors in introducing new food crops in a global economy.

Stuart W. Lewis, M.D.
President and Managing Director
Research Institute for the Study of Man (RISM)
August 2003

Epílogo

Durante los últimos diez años, InterAmericas y el Research Institute for the Study of Man (RISM) han desarrollado y auspiciado en forma conjunta exposiciones y publicaciones, en las que se integran las artes y humanidades con las ciencias sociales. Los archivos de la Fase I de THE NEW OLD WORLD/*El nuevo viejo mundo* no tienen precedentes concretos en las artes o ciencias sociales y constituyen información nueva a partir de la cual los investigadores de diferentes disciplinas podrán extraer nuevos conocimientos. El interés principal de RISM, que publica en forma conjunta con InterAmericas el Tomo IV de *12/12*, CAZABI: *Gift of the Americas*, reside en comprender los factores culturales que rodean al cultivo y procesamiento tradicionales de la yuca, cuya evolución logró eliminar las concentraciones potencialmente tóxicas de glucósidos cianogénicos y proveer carbohidratos para una dieta balanceada en forma eficiente y efectiva.

Los pasajes seleccionados para esta publicación documentan la centralidad de las técnicas de eliminación de la toxicidad de la yuca y de sus combinaciones alimenticias apropiadas dentro de la cultura de los Amerindios del Caribe insular. Cada vez más, los científicos sociales reconocen que la integración de las costumbres y prácticas rituales en la elaboración y consumo de alimentos constituye un mecanismo cultural efectivo de trasmisión de información esencial y necesaria para la supervivencia. La introducción de la yuca en África y en Asia parece haber sido exitosa donde era compatible con las costumbres y prácticas alimentarias de los pueblos indígenas de las Américas. En los casos en los que esto no ocurre, los efectos pueden llegar a ser devastadores, según lo demuestran los brotes de kwashiorkor (enfermedad potencialmente mortal con síntomas múltiples provocados por insuficiencia proteica) o de Konzo (enfermedad de motoneuronas) ocurridos en la República Democrática del Congo, donde el proceso de eliminación de la toxicidad de la yuca se acelera en épocas de sequía o hambruna.

La expansión mundial del cultivo de la yuca sin su contexto cultural concomitante ha generado serias consecuencias sanitarias, entre las que se cuentan enfermedades y desnutrición, ya que una dieta basada en la yuca es deficiente en proteínas y/o yodo, en especial si las raíces no han sido procesadas debidamente. La bioquímica y la genética modernas han demostrado que los glucósidos cianógenos de la yuca, junto con la enzima de origen natural denominada linamarasa, producen el neurotóxico ácido cianhídrico. Un repaso de la literatura médica señala que estos glucósidos afectan la biodisponibilidad de micronutrientes, entre los que se incluyen el yodo y aquellos aminoácidos azufrados metionina y cisteína (que intervienen en el procesamiento del mismo). Si bien la yuca genera uno de los más altos rendimientos de almidón por hectárea, es baja en proteínas, las que a su vez están compuestas básicamente por aminoácidos.

De manera que una dieta en la que sólo se consume yuca provocará desnutrición proteica e inhibición de la capacidad de la glándula tiroides para utilizar yodo. Los Amerindios y sus descendientes actuales consumen yuca y/o casabe junto con pescado, que posee un alto contenido tanto en proteínas como en yodo.

La capacidad para mantener un "desarrollo económico sustentable", esencial para la supervivencia de las poblaciones en expansión de los países en desarrollo, quizá dependa del valor que la ciencia moderna "occidental" le asigne a las prácticas antiguas tanto en la cultura de origen, como en la cultura en la que se introduce el alimento. En vista de ello, tal vez deban de reexa-minarse muchas de las iniciativas de desarrollo presentadas por los europeos o los norteamericanos. RISM fomenta la investigación en las ciencias sociales y la medicina a fin de lograr que el conocimiento tradicional se incorpore a la introducción de nuevos cultivos alimentarios para beneficio y crecimiento mutuos. El revisar la historia de la documentación sobre el "pan de la yuca" puede sentar un precedente en la consideración de los factores culturales al momento de introducir nuevos cultivos alimentarios en una economía global.

Stuart W. Lewis, Doctor en Medicina
Presidente y Director Ejecutivo
Research Institute for the Study of Man (RISM)
Agosto de 2003

Appendix A

Directory of Institutions

Caribbean Contemporary Arts (CCA) is an international arts organization that works with contemporary visual artists, curators, writers, historians, and art educators from the Caribbean and the Caribbean diaspora to exhibit, publish, and document the region's art practice, influences, and ideas. CCA was established to provide a space and structure in response to the social, political, and cultural concerns of the many peoples who comprise the Caribbean Basin, and maintains an active interest in work that investigates and includes a diversity of subjects within contemporary art practice, theory, history, and criticism. In 2000, CCA opened CCA7, Centre for the Contemporary Arts, Trinidad's first multipurpose contemporary arts center, dedicated to planning, education, dialog, exchange, and research in the arts. Charlotte Elias is the Programme and Development Director of CCA.

InterAmericas/Society of Arts and Letters of the Americas/*Sociedad de Artes y Letras de las Américas/ Société des Arts et Lettres des Amériques* is a program of the New York Foundation for the Arts, with offices in the City of New York and Port of Spain, Trinidad, that creates and participates in collaborative projects celebrating the richness and diversity of the arts and humanities of the Americas. Exhibitions that incorporate information-gathering techniques of the social sciences within the context of the contemporary arts are a special interest of InterAmericas. It has been involved with a number of traveling exhibitions, including venues at *Espacio/Espace* InterAmericas Space within CCA7 and the Gallery at the Research Institute for the Study of Man (RISM). CAZABI: *Gift of the Americas* is Volume IV of *12/12*, a twelve-volume commemorative series of portfolios of twelve *giclée* (Iris) prints. The subject of each volume of *12/12* relates to interests and program activities of InterAmericas since its foundation in 1992. Jane Gregory Rubin is the Founder and Director of InterAmericas.

The **Research Institute for the Study of Man (RISM)** was incorporated in 1955 as a non-profit organization for educational and scientific purposes. To implement those goals, RISM has conducted research and training programs, developed an extensive library, supported scholarly exchanges and publications, organized conferences, provided consultation services and stimulated the initiation, development, and dissemination of basic knowledge in the behavioral sciences. From its inception, RISM has fostered and conducted multidisciplinary, cross-cultural studies of contemporary relevance. Basic research projects of applied value have been undertaken in developing and rapidly changing societies, in collaboration with scholars and institutions from the host countries. Much of this seminal work was pioneered in the Caribbean region, which then afforded and continues to provide unique possibilities for understanding the complex interplay of historic, economic, social, and cultural forces in multiethnic societies. This integrated approach to multidisciplinary study now encompasses a broad range of problems and issues in a variety of sociocultural settings in other world areas. Stuart W. Lewis, M.D., is the President and Managing Director of RISM.

George Gustav Heye Center of the Smithsonian National Museum of the American Indian (NMAI). The Smithsonian's National Museum of the American Indian in Lower Manhattan, the George Gustav Heye Center, is located in the historic Alexander Hamilton U.S. Custom House, one of the most splendid Beaux-Arts buildings in New York. The museum features year-round exhibitions about Native American cultures, an extensive film and video program, dance and music performances, children's workshops, as well as a resource and learning center. The Lower Manhattan branch is one of three locations of the Smithsonian's National Museum of the American Indian. Founded by an Act of Congress in 1989 that appropriated funds for the development of facilities at three sites from the former Museum of the American Indian, the museum includes the George Gustav Heye Center in New York City, opened in 1994; the Cultural Resources Center, six miles southeast of the National Mall in Suitland, Maryland, opened in 1999; and the NMAI on the National Mall in Washington, D.C., currently under construction and scheduled to open in September 2004. John Haworth is the Director of the George Gustav Heye Center.

Appendix B

Research Institute for the Study of Man (RISM) Holdings Related to Cassava

Butt, Colson Audrey. "Inter-tribal Trade in the Guiana Highlands." *Antropológía* (Fundaçion La Salle de Ciencias Naturales, Instituto Caribe de Antropología y Sociología, Caracas, Venezuela), no. 34, 1973, pp. 6–70. See esp. pp. 27–34 on "The Cassava Grater (*Chimari*)." [Period covered 1951–1957]

Carneiro, Robert L. "Slash-and-burn Cultivation Among the Kuikuru and Its Implications for Cultural Development in the Amazon Basin." *The Evolution of Horticultural Systems in Native South America: Causes and Consequences: A Symposium*. Johannes Wilbert, ed. Caracas, Venezuela: Sociedad de Ciencias Naturales La Salle, 1961. (*Antropológica* Supplement Publication no. 2, September 1961). pp. 47–69. [Editorial Sucre, Caracas]

Casas, Bartolomé de las. *History of the Indies*. Translated and edited by Andrée Collard. New York: Harper & Row, 1971.

Farrabee, William Curtis. *The Central Caribs*. Oosterhout, N.B. The Netherlands, 1967. (University of Pennsylvania. The University Museum. *Anthropological Publications*, vol. X)

Fernández de Oviedo y Valdés, Gonzalo. *Historia general y natural de las Indias, Islas y Tierra firme del mar Océano*. Edición y estudio preliminar de Juan Pérez de Tudela Bueso. 2d ed. Madrid: Ediciones ATLAS, 1992. 5v. (Biblioteca de Autores Españoles)

____*Natural History of the West Indies*. (*Sumario de la Natural Historia de las Indias*, 1526.) Translated and edited by Sterling A. Stoudemire. Chapel Hill, NC: University of North Carolina Press, 1959. (*Studies in the Romance Languages and Literature* [University of North Carolina], no. 32). pp. 15–19.

Frechione, John. "Manioc Monozoning in Yekuana Agriculture." *Antropológica*, (Fundaçion La Salle de Ciencias Naturales, Instituto Caribe de Antropología y Sociología, Caracas, Venezuela), no. 58, 1982. p. 53–74.

Gerbi, Antonello. *Nature in the New World*. (*La natura delle Indie nove*, 1975.) Translated by Jeremy Moyle. University of Pittsburgh Press, Pittsburgh, PA, 1985.

Hulme, Peter and Neil L. Whitehead, eds. *Wild Majesty: Encounters with Caribs from Columbus to the Present day*, an anthology. Oxford, England: Clarendon Press, 1992.

Joyce, Thomas A. *Central American and West Indian Archaeology; Being an Introduction to the Archaeology of the States of Nicaragua, Costa Rica, Panama and the West Indies*. London: Philip Lee Warner, 1916.

Keegan, William F. *The People Who Discovered Columbus: The Prehistory of the Bahamas*. Gainesville, FL: University Press of Florida, 1992.

Leeds, Anthony. "Yaruro Incipient Tropical Forest Horticulture—Possibilities and Limits." *The Evolution of Horticultural Systems in Native South America: Causes and Consequences: A Symposium*. Johannes Wilbert, ed. Caracas, Venezuela: Sociedad de Ciencias Naturales La Salle, 1961. (*Antropológica* Supplement Publication, no. 2, September 1961), pp. 13–46.

Lovén, Sven. *Origins of the Tainan culture, West Indies*. Göteborg, Sweden: Elanders Bokfryckeri Akfiebolag, 1935, pp. 358–368.

Mentore, George P. "Wai-wai Labour Relations in the Production of Cassava." *Antropológica* (Fundaçion La Salle de Ciencias Naturales, Instituto Caribe de Antropología y Sociología, Caracas, Venezuela), no. 59–62, 1983–1984, pp. 199–221.

Olsen, Fred. *On the Trail of the Arawaks.* Norman, OK: University of Oklahoma Press, 1974. (The Civilization of the American Indian Series, vol. 129)

Petitjean Roget, Jacques. "A propos des platines à manioc." *Proceedings of the Sixth International Congress for the Study of Pre-Columbian Cultures of the Lesser Antilles, Pointe-a-Pitre, Guadeloupe, July 7–12, 1975.* Gainesville, FL: Ripley P. Bullen and R. Gaskin, 1976. pp. 76–81.

Simon, O. R. "Supporting Evidence for the Use of Cassava (*Manihot esculenta*) Products Instead of Wheat Flour Products in the Diet of Diabetics." *West Indian Medical Journal,* vol. 37, no. 2, June 1988. pp. 100–105.

Steward, Julian Haynes. *Handbook of South American Indians.* Washington, DC: United States Government Printing Office, 1959. 7v. (There are many many pages devoted to discussions of cassava and manioc throughout the seven volumes. Volume 7, the index to the previous six volumes, references every page in every volume.)

Sturtevant, William C. "Taino Agriculture." *The Evolution of Horticultural Systems in Native South America: Causes and Consequences: A Symposium.* Johannes Wilbert, ed. Caracas, Venezuela: Sociedad de Ciencias Naturales La Salle, 1961. (*Antropológica* Supplement Publication, no. 2, September 1961). pp. 69–82. [Editorial Sucre, Caracas]

____"Taino agriculture (1961)." *Earliest Hispanic/Native American Interactions in the Caribbean.* Edited with an introduction by William F. Keegan. (Spanish Borderlands Source Books; 13) New York: Garland Publishing, 1991. pp. 241–255.

Taylor, Douglas. *The Caribs of Dominica.* Washington, D.C.: United States Government Printing Office, 1938. (Smithsonian Institution. Bureau of American Ethnology. Anthropological Papers, No. 3).

Walker, D. J. R. *Columbus and the Golden World of the Arawaks: The Story of the First Americans and Their Caribbean Environment.* Kingston: Ian Randle, 1992.

Watts, David. *The West Indies: Patterns of Development, Culture and Environmental Change Since 1492.* Cambridge: Cambridge University Press, 1987.

Wilbert, Johannes, Ed. *The Evolution of Horticultural Systems in Native South America: Causes and Consequences: A Symposium.* Caracas, Venezuela: Sociedad de Ciencias Naturales La Salle, 1961. [Editorial Sucre, Caracas] (*Antropológica* Supplement Publication no. 2, September 1961). See individual items under Carneiro, Leeds, and Sturtevant.

Wilson, Samuel M. and G. London. *Hispaniola: Caribbean Chiefdoms in the Age of Columbus.* Tuscaloosa, Ala: University of Alabama Press, 1990. References to cassava and manioc.

Yde, Jens. *Material Culture of the Waiwái.* Copenhagen: [Nationalmuseets Skrifter Etnografisk Rœkke, X] The National Museum of Copenhagen, 1965. See esp. pp. 23–53.

Appendix C

Photographic Credits

From CAZABI: *Gift of the Americas*, Volume IV of *12/12*

PLATE 5 *frontispiece*
08.16.2001, BARRANQUITAS

PLATE 11 *page* 20
02.01.2000, ARIMA

PLATE 7 *page* 38
02.01.2000, ARIMA

PLATE 3 *page* 68
06.15.2002, ISABELA

PLATE 4 *page* 96
06.16.2002, ISABELA

PLATE 9 *page* 112
06.16.2002, ISABELA

Courtesy of James Pepper Henry

05.21.2000 CARIDAD DE LOS INDIOS *page* 40
Marisol Villanueva

From THE NEW OLD WORLD/*El nuevo viejo mundo*: Selections from the archives of Phase I

07.17.1999 BARRANQUITAS *page* 45

07.17.1999 BARRANQUITAS *page* 46
Nina M. Raffaele Aponte

07.17.1999 BARRANQUITAS *page* 47

Notes/Notas

Historia general y natural de las Indias, Islas y Tierra firme del mar Océano

Editor's Note

1. Sterling A. Stoudemire, *Natural History of the West Indies.* (*Sumario de la natural historia de las Indias*, 1526.) University of North Carolina Press, Chapel Hill, NC, 1959 (*Studies in the Romance Languages and Literature* [University of North Carolina], No. 32), ix.

2. Antonello Gerbi, *Nature in the New World*, translated by Jeremy Moyle from *La natura delle Indie nove* (1975). University of Pittsburgh Press, Pittsburgh, PA, 1985, 131.

3. Ibid, 129; see also Stoudemire, xvii.

4. Stoudemire, xvii.

5. Gerbi, 130.

6. Gerbi, 133.

Part I, Book VII, Chapter 2

1. An arroba is an old Spanish unit of weight equal to about 25 pounds.

Part II, Book XXIV, Chapter 3

1. The river referred to is the Orinoco, and the region described is the Orinoco River Valley, across the Gulf of Paria from the western coast of Trinidad.

2. Here, Fernández de Oviedo's footnote reads "*Ethim.*, book XIII, chap. XXI and book XII, chap. VI," to refer to *Etymologiae sive Originum libri* XX, also known as *Origenes*, by St. Isidore of Seville (c.560–636 CE).

3. Here, Fernández de Oviedo's footnote reads "Book V, chapter X" to refer to the 37-volume work *Natural History*, by Pliny the Elder, or Caius Plinius Secundus (c.23–79 CE).

Nota del Editor

1. Sterling A. Stoudemire, *Natural History of the West Indies.* (*Sumario de la natural historia de las Indias*, 1526.) Chapel Hill, NC: University of North Carolina Press, 1959 (*Studies in the Romance Languages and Literature* [University of North Carolina], No. 32), ix.

2. Antonello Gerbi, *Nature in the New World*, traducción de *La natura delle Indie nove* (1975) por Jeremy Moyle. University of Pittsburgh Press, Pittsburgh, PA, 1985, 131.

3. Ibíd, 129; ver también Stoudemire, xvii.

4. Stoudemire, xvii.

5. Gerbi, 130.

6. Gerbi, 133.

Primera parte, Libro VII, Capítulo 2

1. Una arroba es una antigua medida de peso española que equivale a alrededor de 25 libras.

Segunda parte, Libro XXIV, Capítulo 3

1. El río al que se hace referencia es el Orinoco, y la región descripta es el Valle del Río Orinoco que, visto desde la costa occidental de Trinidad, se encuentra situado del otro lado del Golfo de Paria.

2. Aquí la nota a pie de página de Fernández de Oviedo dice "*Ethim.*, lib. XIII, cap. XXI, y lib. XII, cap. VI." Se refiere a la obra de San Isidro de Sevilla (c.560–636 CE) *Etymologiae sive Originum libri* XX, también conocida como *Origenes*.

3. Aquí la nota a pie de página de Fernández de Oviedo dice "Lib. V, cap. X." Se refiere a la obra en 37 volúmenes de la *Historia Natural* escrita por Plinio El Viejo o Cayo Plinio Segundo (c.23–79 CE).

Resistencia y supervivencia indígena en Puerto Rico/Survival of *la cultura de la yuca* in Puerto Rico

1. Delgado, Juan Manuel, *Mártires de la Nación Puertorriqueña*, Biblioteca de Historia Nacional, Florida, Puerto Rico, 2002.

2. Gelpí Baíz, Elsa, *Siglo en blanco: estudio de la economía azucarera en el Puerto Rico del siglo XVI (1540–1612)*, Editorial de la Universidad de Puerto Rico, San Juan, Puerto Rico, 1999, 42–43.

3. Canino Salgado, Marcelino, *Gozos devocionales de la tradición puertorriqueña*; Editorial Universitaria, Universidad de Puerto Rico, Río Piedras, Puerto Rico, 1974, 11.

4. Vidal, Teodoro, "Aportación al estudio del folklore médico en Puerto Rico," *Revista del Instituto de Cultura Puertorriqueña*, Number 50 (January–March 1971), 63.

5. Vansina, Jan, *De la tradition orale; essai de méthode historique*, Tervuren, 1961.

6. López Cantos, Angel, *La religiosidad popular en Puerto Rico (siglo XVIII)*, Centro de Estudios Avanzados de Puerto Rico y el Caribe, San Juan, Puerto Rico, 1993, 42.

7. Abbad y Lasierra, Iñigo, *Historia geográfica, civil y natural de la isla de San Juan Bautista de Puerto Rico*, Ediciones de la Universidad de Puerto Rico, Río Piedras, Puerto Rico, 1959, 193; quoted by López Cantos, 42.

8. Campo Lacasa, Cristina, *Historia de la Iglesia en Puerto Rico, 1511–1802*, Instituto de Cultura Puertorriqueña, San Juan, Puerto Rico, 1977, 145.

9. Lamourt-Valentín, Oscar, *La vierge au Rosaire*, of his series *Cuadernos de estudios nativos*, Centro Graáfico del Caribe, Inc., Puerto Rico, 1992.

10. Cabanillas de Rodríguez, Berta, *El folklore en la alimentación puertorriqueña*, Editorial de la Universidad de Puerto Rico, Río Piedras, Puerto Rico, 1983, 50–51 and 54.

11. Pompa, Gerónimo, *Medicamentos Indígenas*, Libros EASA, Madrid, 1979, 28, 71, 254–255.

12. Coll y Toste, Cayetano, *Prehistoria de Puerto Rico*, Editorial Vasco Americana, s. f., Bilbao, 1970, 96–97.

13. Alvárez Nazario, Manuel, *El arcaísmo vulgar en el español de Puerto Rico*, Mayaguez, Puerto Rico, 1957, 173.

14. Letter sent by the Bishop of Puerto Rico, Friar Damián López de Haro, to Juan Díaz de la Calle, September 27, 1644.

15. Babín, María Teresa, "Símbolos de Borinquen;" *Revista del Instituto de Cultura Puertorriqueña*, No. 4, July–September, 1959, 2.

Resistencia y supervivencia indígena en Puerto Rico/Sobrevivencia de la cultura de la yuca en Puerto Rico

1. Delgado, Juan Manuel, *Mártires de la Nación Puertorriqueña*, Biblioteca de Historia Nacional, Florida, Puerto Rico, 2002.

2. Gelpí Baíz, Elsa, *Siglo en blanco: estudio de la economía azucarera en el Puerto Rico del siglo XVI (1540–1612)*, Editorial de la Universidad de Puerto Rico, San Juan, Puerto Rico, 1999, 42–43.

3. Canino Salgado, Marcelino, *Gozos devocionales de la tradición puertorriqueña*; Editorial Universitaria, Universidad de Puerto Rico, Río Piedras, Puerto Rico, 1974, 11.

4. Vidal, Teodoro, "Aportación al estudio del folklore médico en Puerto Rico", *Revista del Instituto de Cultura Puertorriqueña*, Número 50, enero–marzo, 1971, 63.

5. Vansina, Jan, *De la tradition orale; essai de méthode historique*, Tervuren, 1961.

6. López Cantos, Angel, *La religiosidad popular en Puerto Rico (siglo XVIII)*, Centro de Estudios Avanzados de Puerto Rico y el Caribe, San Juan, Puerto Rico, 1993, 42.

7. Abbad y Lasierra, Iñigo, *Historia geográfica, civil y natural de la isla de San Juan Bautista de Puerto Rico*, Ediciones de la Universidad de Puerto Rico, Río Piedras, Puerto Rico, 1959, 193; citado por López Cantos, 42.

8. Campo Lacasa, Cristina, *Historia de la Iglesia en Puerto Rico, 1511–1802*, Instituto de Cultura Puertorriqueña, San Juan, Puerto Rico, 1977, 145.

9. Lamourt-Valentín, Oscar, *La vierge au Rosaire*, de su serie *Cuadernos de estudios nativos*, Centro Graáfico del Caribe, Inc., Puerto Rico, 1992.

10. Cabanillas de Rodríguez, Berta, *El folklore en la alimentación puertorriqueña*, Editorial de la Universidad de Puerto Rico, Río Piedras, Puerto Rico, 1983, 50–51 y 54.

11. Pompa, Gerónimo, *Medicamentos Indígenas*, Libros EASA, Madrid, 1979, 28, 71, 254–255.

12. Coll y Toste, Cayetano, *Prehistoria de Puerto Rico*, Editorial Vasco Americana, s. f., Bilbao, 1970, 96–97.

13. Alvárez Nazario, Manuel, *El arcaísmo vulgar en el español de Puerto Rico*, Mayaguez, Puerto Rico, 1957, 173.

14. Carta del Obispo de Puerto Rico, Fray Damián López de Haro, a Juan Díaz de la Calle, 27 de septiembre de 1644.

15. Babín, María Teresa, "Símbolos de Borinquen", *Revista del Instituto de Cultura Puertorriqueña*, Número 4, julio–septiembre, 1959, 2.

Manihot Esculenta: Historic and Economic Context

1. *The Log of Christopher Columbus*, translated by Robert H. Fuson (Camden, Maine: International Marine Publishing, 1992).

2. Lucio Sorre, "His Voyages and Discovery of the New World," in "Christopher Columbus: His Gastronomic Persona." Article published online by Castellobanfi restaurant and winery, Montalcino, Italy, 1997. See www.castellobanfi.com/features/story_3.html.

3. The 1997–98 exhibition "A Harvest Gathered: Food in the New World," at the John Carter Brown Library at Brown University in Providence, Rhode Island, highlighted that nearly 30 percent of the world's cultivated plants came from the New World. Cassava is one of these plants that spread through the Colombian exchange.

4. The hydrocyanic acid (HCN) in cassava can cause asphyxiation if not properly processed. The lethal concentration of HCN is 150 milligrams for a 50-kilogram adult. See "Writeups and Illustrations of Economically Important Plants." Essay published online for the economic botany course "Plants and Civilization," University of California at Los Angeles, taught by Dr. Arthur C. Gibson. See www.botgard.ucla.edu/html/botanytextbooks/economicbotany/Manihot/index.html.

5. Samuel Turner, "The Conquest of Higüey: The Eyewitness Account of Las Casas Examined and the Archaeological Implications for the Parque Nacional del Este, República Dominicana." Article published online by the Dominican Republic Research Underwater Science Program, Indiana University. See www.indiana.edu/~r317 doc/dr/higuey/higuey3.html.

6. Fuson, *The Log of Christopher Columbus*. See Appendix F, "Roots and Tubers," 235.

7. Turner, "The Conquest of Higüey," 2.

8. Martha Groves, "Plant Researchers Offer Bumper Crop of Humanity," *Los Angeles Times*, December 25, 1997, 1. Part of investigative series "Far From Plenty: Biotechnology's Promise for Feeding."

9. S.A. Balogun, J.B. Bodin, N. Bikangi, I. Rafiqul and I. Jarlebring, "Cassava: The Ultimate Future Crop." Article published online by the Global Nutrition Project, Department of Medical Sciences, Uppsala University, Uppsala, Sweden, September 7, 1998, 2. See www.nutrition.uu.se/studentprojects/group97/cassava/cassava.html.

10. On October 30, 2002, the World Health Organization published a report citing the ten major health risks worldwide that account for 40 percent of 56 million deaths

each year. Lack of food was number one, causing 3.4 million deaths in 2000. See Agence France-Presse, "Agency Puts Hunger No. 1 on List of World's Top Health Risks," *The New York Times*, October 31, 2002.

11. Dolores R. Piperno, Anthony J. Ranere, Irene Holst, and Patricia Hansell, "Starch Grains Reveal Early Root Crop Horticulture in the Panamanian Tropical Forest," *Nature* 407 (2000): 894–897.

12. Ibid.

13. John Whitfield, "Roots of Farming," *Nature News Service* (October 19, 2000): 1.

14. Mary Karasch in *The Cambridge World History of Food* writes that other scholars think there could be two regions of origin: Brazil and Central America. Still others hypothesize that one type of cassava (bitter) may have been domesticated in South America, while the sweet variety was domesticated in Central America. See "Manioc" in vol. 1 (Cambridge, U.K., and New York: Cambridge University Press, 2000), 182.

15. Chelsie Vandaveer, "What Common Food Plant Was Used for Hunting?" Article published online under *Herbal Folklore Update*, sponsored by *National Geographic* (October 15, 2001). See www.killerplants.com/herbal-folklore/20011015.asp.

16. The *Yákwa* is the most important annual ritual of the Enauênê-Nauê people of Brazil. During this seven-month ritual, the Enauênê-Nauê plant cassava root in their collective fields. See the documentary film *Yákwa: The Banquet of the Spirits*, directed by Virginia Valadão, Altair Paixão, and Vincent Carelli (1995). This film was shown in 2002 at The British Museum, London, in conjunction with the exhibition "Unknown Amazon: Culture in Nature in Ancient Brazil."

17. Mark J. Plotkin, *Tales of a Shaman's Apprentice, An Ethnobotanist Searches for New Medicines in the Amazon Rain Forest* (New York: Penguin Books, 1993), 107.

18. Ibid.

19. Fuson, *The Log of Christopher Columbus*, 137.

20. Ibid.

21. Samuel M. Wilson, "Columbus, My Enemy: A Caribbean Chief Resists the First Spanish Invaders," *Natural History* (December 1990): 2.

22. Adolfo López Belando, "*El Contacto de los Aborigenes Antillanos con los Colonizadores Españoles Documentado en las Pinturas de la Cueva de 'José María.'*" Article published online by the Dominican Republic Research Underwater Science Program, Indiana University, April 5, 1997. See www.indiana.edu/~r317doc/dr/adolfo/adolfo.html.

23. John W. Foster and Adolfo López Belando, "Images of Encounter and Tribute: Early Sixteenth Century Pictographs from the Jose Maria Cave." Paper presented at the annual meeting of the Society of Historical Archaeology, Corpus Christi, Texas, January 10, 1997.

24. "The Taíno Indians and the Jose Maria Cave." Article published online by the Underwater Science Program, Indiana University, October 14, 1996, 1. See www.indiana.edu/~r317doc/dr/cavehist.html.

25. See Dr. Lynne Guitar, doctoral thesis, "Cultural Genesis: Relationships among Indians, Africans, and Spaniards in Rural Hispaniola, First Half of the Sixteenth Century," Vanderbilt University, 1998. Guitar writes in Chapter 2, 64: "Columbus levied a quarterly tribute of one gold-filled hawk's bell on each Taíno over age fourteen, to be collected by the *caciques* (a hawk's bell was about the size of today's average 'jingle' bell). Later he agreed to accept food and cotton textiles in tribute, for the Taínos were unable to supply so much gold...."

26. Foster and López Belando, "Images of Encounter and Tribute: Early Sixteenth Century Pictographs from the Jose Maria Cave."

27. "The Taíno Indians and the Jose Maria Cave." Op. cit.

28. Turner, "The Conquest of Higüey," 1.

29. The sixteenth-century Spanish historian Fray Bartolomé de las Casas gives the only existing eyewitness account. Quoted in Turner, "The Conquest of Higüey." See 2 for a summary of De las Casas' horrified account. While the Taíno loaded the bread onto the ship, their *cacique* stood by, supervising and directing the Taíno with his scepter. Nearby, an attack dog on a chain lunged at the Taíno chieftain and disemboweled him as his people looked on. Turner writes: "The news of the incident quickly spread throughout Higüey. The *cacique* Cotubanamá was outraged and promptly armed himself and his warriors...."

30. Ibid.

31. Ibid. According to Turner, this violated an agreement that the Spanish captain, General Juan de Esquivel, had made with the Taínos. Esquivel and his men later hunted down and captured Cotubanamá, and Esquivel turned the last free Taíno *cacique* over to the Spanish governor for hanging.

32. Turner, "The Conquest of Higüey," 8.

33. See Guitar, doctoral thesis, 3: "Although there are a number of people today on Hispaniola, Puerto Rico and Cuba, and others who have emigrated to the United States, who are aware that they are descendants of the Taínos, they have lost much of their Classic history, language, customs and traditions. Many others, also Taíno descendants, are totally unaware of the indigenous inheritance. This is because the Taínos who managed to survive the wars, abuse, famines and multiple epidemics after 1492 by fleeing to peripheral regions, led isolated lives that were far different from the well-organized, intensively agricultural, well-populated societies first encountered by the Europeans. Others survived by intermarrying with Spaniards, with natives imported to the Greater Antilles from other regions of the Americas, or with Africans. Many Taíno socio-cultural traditions were subsumed under the barrage of customs, beliefs and traditions that arrived with the new voluntary and involuntary immigrants, whose socio-cultural patterns and traditions were adopted and intermixed with Taíno customs and traditions by their mixed-blood *criollo* ('born in the Americas') descendants."

34. See the official government tourism Web site of the Dominican Republic: www.terra.com.do/turismodo/lacotica/es/gastronomia/gastronomia.html.

35. S.E. Carter, L.O. Fresco, G. Jones, and J.N. Fairbairn, "Introduction and Diffusion of Cassava in Africa," *IITA Research Guide* 49 (July 1997): 1.

36. S.E. Carter, et al., "Initial Introduction" *IITA Research Guide* 49 (July 1997).

37. S.E. Carter, et al., "Central Africa" section, *IITA Research Guide* 49 (July 1997).

38. Ibid.

39. S.E. Carter, et al., "West Africa" section, *IITA Research Guide* 49 (July 1997).

40. Taye Babaleye of the International Institute of Tropical Agriculture (IITA) writes that "[n]ew knowledge of the biochemistry of the crop has proved that the proteins embedded in the leaves are equal in quality to the protein in egg...." Taye Babaleye, "Cassava, Africa's Food Security Crop." Online newsletter of Consultative Group on International Agricultural Research (CGIAR), Vol. 3, No. 1 (March 1996). See also www.nutrition.uu.se/studentprojects/group97/cassava/casnut.html: The leaves are carefully chosen and then pounded in a mortar or cut up with a knife. The pounded leaves are then boiled for an hour. Cooks add oil, hot pepper, and salt along with ground nuts, sesame, elephant grass, seafood or meat.

41. Ari Nave, "Mauritius." Article published by the online magazine *Black World: News & Views*. See www.africana.com/research/encarta/tt_381.asp.

42. Mary Karasch, "Manioc," *The Cambridge World History of Food*, vol. 1 (Cambridge, UK, and New York: Cambridge University Press, 2000), 183–4.

43. "History of Bogor Botanic Garden." Article published online by the Indonesian Network for Plant Conservation, January 2003. See www.bogor.indo.net.id/ kri/bhist.html.

44. See *Processing of Cassava*, Agriculture Services Bulletin No. 8, United Nations Food and Agriculture Organization (Rome: FAO, 1971). See www.fao.org /inpho/ vlibrary/x0032e/X0032E01.html.

45. International Center for Tropical Agriculture (CIAT), "Cassava Boom in Southeast Asia." Online newsletter of Consultative Group on International Agricultural Research (CGIAR), Vol. 4, No. 3, (June 1997). See www.worldbank.org/html/cgiar/newsletter/ june97/9ciat.html.

46. Klanarong Sriroth, Kuakoon Piyachomwan, Kunruedee Sangseethong, and Christopher Oates, "Modification of Cassava Starch." Paper presented at the 10th International Starch Convention, Cracow, Poland, June 11–14, 2002. See www.cassava.org/Poland/Modification.pdf.

47. See "Improved Germplasm," a Web site of the International Center for Tropical Agriculture (CIAT), Bogotá, Colombia. See www.ciat.cgiar.org/improved_germ plasm/cassava.htm, 2001.

48. "Championing the Cause of Cassava," *News & Highlights*, online publication of the Food and Agriculture Organization of the United Nations (FAO), April 26, 2000, 1. See www.fao.org/news/2000/000405-e.html.

49. CIAT agronomist Reinhardt Howeler, quoted in "Cassava in Southeast Asia: From Hard Times to Modern Times." Op. cit.

50. "What to do About Whitefly?" Article published online in *New Agriculturist on-line: Reporting Agriculture for the 21st Century* (October 2002). See www.new-agri.co.uk/02-3/newsbr.html. See also Mike Crawley, "Restoring Cassava Production in Uganda." Article published online in *REPORTS: Science from the Developing World* (Ottawa, Canada: International Development Research Centre (IDRC), November 12, 1999. See www.idrc.ca/ reports/read_article_english.cfm?article_num=558.

51. Myambu Banea, Nigel H. Poulter, and Hans Rosling, "Shortcuts in Cassava Processing and Risk of Dietary Cyanide Exposure in Zaire." Article published online in *Food and Nutrition Bulletin* 14:2 (June 1992) (Tokyo: United Nations University). See www.unu.edu/ unupress/food/8F142Eoa.htm.

52. The Cassava Cyanide Disease Network based at the School of Botany and Zoology of the Australian National University in Canberra, Australia, monitors poisoning, and links scientists in several countries. "The network has warned that inadequate processing and over-dependence on cassava has led to widespread poisoning in parts of Africa....Traditional methods of processing practiced in East Africa, such as sun-drying and heap fermentation, leave large quantities of the poison in the cassava flour, in contrast to the roasting method used in west Africa, which removes most of the cyanide. It has also recommended support for a greater diversity of staple crops in cassava dependent areas." In "Fresh Evidence in Cassava Poisoning." Article published online by *New Agriculturist on-line: Reporting Agriculture for the 21st Century*.

53. Average annual world prices for cassava pellets began falling steadily in 1996 and continued to decline through 2000 because of competitive grain pricing and lower demand on the part of the European Union. From: "Cassava in Southeast Asia: From Hard Times to Modern Times." Op. cit.

54. "Championing the Cause of Cassava." Op. cit.

Manihot esculenta: Su contexto histórico y económico

1. *The Log of Christopher Columbus*, traducido por Robert H. Fuson (Camden, Maine: International Marine Publishing, 1992).

2. Lucio Sorre, "His Voyages and Discovery of the New World", en "Christopher Columbus: His Gastronomic Persona". Artículo publicado *online* por el restaurante y bodega Castellobanfi, Montalcino, Italia, 1997. Ver www.castellobanfi.com/features/story_3.html.

3. La exposición de 1997–98 "A Harvest Gathered: Food in the New World", realizada en la Biblioteca John Carter Brown de la Universidad Brown en Providence, Rhode Island, destacó el hecho de que casi el 30 por cientos de las plantas cultivadas en el mundo provenían del nuevo mundo. La yuca es sólo una de dichas plantas que se diseminaron a través del intercambio iniciado por Colón.

4. El ácido cianhídrico (HCN) presente en la yuca puede provocar asfixia si no se lo procesa debidamente. La concentración mortal de HCN es de 150 miligramos en un adulto de 50 kilos. Ver "Writeups and Illustrations of Economically Important Plants". Ensayo publicado en el sitio Web del curso de botánica económica "Plants and Civilization" de la Universidad de California en Los Angeles a cargo del Dr. Arthur C. Gibson. Ver www.botgard.ucla.edu/html/botanytextbooks/economicbotany/ Manihot/index.html.

5. Samuel Turner, "The Conquest of Higüey: The Eyewitness Account of Las Casas Examined and the Archaeological Implications for the Parque Nacional del Este, República Dominicana". Artículo publicado *online* por el Dominican Republic Research Underwater Science Program, Universidad de Indiana. Ver www.indiana.edu/~r317doc/dr/higuey/higuey3.html.

6. Fuson, *The Log of Christopher Columbus*. Ver Apéndice F, "Roots and Tubers", 235.

7. Turner, "The Conquest of Higüey", 2.

8. Martha Groves, "Plant Researchers Offer Bumper Crop of Humanity", *Los Angeles Times*, 25 de diciembre de 1997, 1. Parte de una serie de investigación titulada "Far From Plenty: Biotechnology's Promise for Feeding".

9. S.A. Balogun, J.B. Bodin, N. Bikangi, I Rafiqul y I. Jarlebring, "Cassava: The Ultimate Future Crop". Artículo publicado online por Global Nutrition Project, Departamento de Ciencias Médicas, Universidad de Upsala, Upsala, Suecia, 7 de septiembre de 1998, 2. Ver www.nutrition.uu.se/studentprojects/group97/cassava/cassava/html.

10. El 30 de octubre de 2002, la Organización Mundial de la Salud publicó un informe en el que citaba los diez principales riesgos para la salud en todo el mundo que representan el 40 por ciento de los 56 millones de muertes que se producen cada año. La falta de alimentos figuraba en primer lugar, causando 3,4 millones de muertes en 2000. Ver Agence France-Presse, "Agency Puts Hunger No. 1 on List of World's Top Health Risks". *The New York Times*, 31 de octubre de 2002. Ver www.nytimes.com/2002/10/31/international/31HEAL-FOR.html.

11. Dolores R. Piperno, Anthony J. Ranere, Irene Holst, y Patricia Hansell, "Starch Grains Reveal Early Root Crop Horticulture in the Panamanian Tropical Forest", *Nature* 407 (2000): 894–897.

12. Ibíd.

13. John Whitfield, "Roots of Farming", *Nature News Service* (19 de octubre de 2000): 1.

14. Mary Karasch en *The Cambridge World History of Food* escribe que otros académicos sostienen que podrían existir dos regiones de origen: Brasil y América Central. Otros aún apoyan la hipótesis de que un tipo de mandioca (amarga) puede haberse aclimatado en América del Sur mientras que la variedad dulce se aclimató en América Central. Ver "Manioc" en vol. 1 (Cambridge, GB, y Nueva York: Cambridge University Press, 2000), 182.

15. Chelsie Vandaveer, "What Common Food Plant Was Used for Hunting?" Artículo publicado *online* en *Herbal Folklore Update*, auspiciado por *National Geographic* (15 de octubre de 2001). Ver www.killerplants.com/herbal-folklore/20011015.asp.

16. El *Yákwa* es el ritual anual más importante de los Enauênê-Nauê del Brasil. Durante los siete meses que dura este ritual, los Enauênê-Nauê siembran la yuca en sus campos colectivos. Ver el documental *Yákwa: The Banquet of the Spirits*, dirigido por Virginia Valadão, Altair Paixão, y Vincent Carelli (1995). Este documental fue exhibido en 2002 en el British Museum, Londres, conjuntamente con la exposición "Unknown Amazon: Culture in Nature in Ancient Brazil".

17. Mark J. Plotkin, *Tales of a Shaman's Apprentice, An Ethnobotanist Searches for New Medicines in the Amazon Rain Forest* (Nueva York: Penguin Books, 1993): 107.

18. Ibíd.

19. Samuel M. Wilson, "Columbus, My Enemy: A Caribbean Chief Resists the First Spanish Invaders", *Natural History* (diciembre de 1990): 137.

20. Adolfo López Belando, "El Contacto de los Aborígenes Antillanos con los Colonizadores Españoles Documentado en las Pinturas de la Cueva de 'José María'". Artículo publicado *online* por el Dominican Republic Research, Underwater Science Program, Universidad de Indiana, 5 de abril de 1997. Ver www.indiana.edu/~r317doc/dr/adolfo/adolfo.html.

21. John W. Foster y Adolfo López Belando, "Images of Encounter and Tribute: Early Sixteenth Century Pictographs from the Jose Maria Cave". Trabajo presentado en la reunión anual de la Sociedad de Arqueología Histórica, Corpus Christi, Texas, 10 de enero de 1997.

22. "The Taíno Indians and the José María Cave". Artículo publicado *online* en Dominican Republic Research, Underwater Science Program, Universidad de Indiana, 14 de octubre de 1996, 1. Ver www.indiana.edu/~r317doc/dr/cavehist.html.

23. Ver Dr. Lynne Guitar, tesis doctoral, "Cultural Genesis: Relationships among Indians, Africans, and Spaniards in Rural Hispaniola, First Half of the Sixteenth Century". Universidad de Vanderbilt, 1998. Guitar escribe en el capítulo 2, 64: "Colón recaudó un impuesto trimestral equivalente a un cascabel de gavilán lleno de oro por cada taíno de más de catorce años, a ser recolectado por los *caciques* (un cascabel de gavilán equivalía aproximadamente en tamaño a un cascabel actual). Más tarde aceptó comida y telas de algodón en concepto de tributo, ya que los taínos no pudieron proveer de tanto oro...."

24. Foster y López Belando, "Images of Encounter and Tribute: Early Sixteenth Century Pictographs from the Jose Maria Cave".

25. "The Taíno Indians and the José María Cave".

26. Turner, "The Conquest of Higüey", 1.

27. El historiador español del siglo XVI Fray Bartolomé de las Casas proporciona el único relato testimonial existente. Citado en Turner, "The Conquest of Higüey". Ver página 2 para un resumen del relato de horror por De las Casas. Mientras los taínos cargaban el pan al barco, su *cacique* estaba parado supervisando y agitando su cetro. Un perro de ataque español que se encontraba encadenado cerca, arremetió contra el jefe taíno y destripó al líder taíno mientras su gente miraba, según De las Casas. Turner escribe: "...La noticia del incidente pronto se extendió por todo Higüey. El *cacique* Cotubanamá estaba furioso y sin demora alguna se armó a sí mismo y a sus guerreros...."

28. Ibíd.

29. Ibíd. De acuerdo a Turner, ésto violó un acuerdo que el capitán español, General Juan de Esquivel, había hecho con los taínos. Esquivel y sus hombres posteriormente salieron a la búsqueda y capturaron a Cotubanamá. Esquivel entregó el último *cacique* taíno libre al gobernador español para que fuera ahorcado.

30. Turner, "The Conquest of Higüey", 8.

31. Ver Guitar, tesis doctoral, 3: "Aunque hoy existen un número de personas en Hispaniola, Puerto Rico, y Cuba y

otros que han emigrado a los Estados Unidos que saben que descienden de taínos, los mismos han perdido mucho de su historia, idioma, costumbres y tradiciones clásicas. Muchos otros, también descendientes de taínos, desconocen su herencia indígena. Esto obedece a que aquellos taínos que lograron sobrevivir después de 1492 las guerras, los abusos, las hambrunas y las múltiples epidemias al escapar a regiones periféricas, llevaron una vida aislada, muy diferente de aquellas sociedades bien organizadas, con agricultura intensiva y muy pobladas que los europeos encontraron al principio. Otros sobrevivieron casándose con españolas, con nativos importados a las Antillas Mayores procedentes de otras regiones de las Américas o con africanos. Muchas tradiciones socioculturales de los taínos quedaron subsumidas bajo el aluvión de costumbres, creencias y tradiciones que llegaron con los nuevos inmigrantes voluntarios e involuntarios, cuyos patrones y tradiciones socioculturales fueron adoptados y entremezclados con las costumbres taínas por sus descendientes *criollos* ('nacidos en las Américas') de sangre mixta".

32. Ver el sitio Web oficial de turismo del gobierno de República Dominicana: www.terra.com.do/turismodo/lacotica/es/gastronomia/gastronomia/html.

33. S.E. Carter, L.O. Fresco, P.G. Jones, y J.N. Fairbairn, "Introduction and Diffusion of Cassava in Africa", *IITA Research Guide* 49 (julio de 1997): 1.

34. S.E. Carter y otros, "Initial Introduction", *IITA Research Guide* 49 (julio de 1997).

35. S.E. Carter y otros, Sección "Central Africa", *IITA Research Guide* 49 (julio de 1997).

36. Ibíd.

37. S.E. Carter y otros, Sección "West Africa", *IITA Research Guide* 49 (julio de 1997).

38. Taye Babaleye, del Instituto Internacional de Agricultura Tropical (IITA por sus siglas en inglés), escribe "[n]uevos conocimientos acerca de la bioquímica del cultivo han comprobado que las proteínas contenidas en las hojas igualan en calidad a la proteína del huevo...." Taye Babaleye, "Cassava, Africa's Food Security Crop": Boletín *online* del Grupo Consultivo sobre Investigación Agrícola Internacional (GCIAI), Vol. 3, No. 1 (marzo de 1996). Ver también www.nutrition.uu.se/studentprojects/group97/cassava/casnut.html: Se seleccionan las hojas con cuidado y se las machaca en un mortero o se las corta a cuchillo. Luego se hierven las hojas machacadas durante una hora. Los cocineros agregan aceite, pimiento picante, y sal junto con nueces molidas, sésamo, hierba elefante (Pennisetum), mariscos, o carne.

39. Ari Nave, "Mauritius". Artículo publicado en la revista online *Black World: News & Views*. Ver www.africana.com/Articles/tt_381.html.

40. Mary Karasch, "Manioc", *The Cambridge World History of Food*, vol. 1 (Cambridge, GB, y Nueva York: Cambridge University Press, 2000), 183–4.

41. "History of Bogor Botanic Garden". Artículo en el sitio Web de Indonesian Network for Plant Conservation, enero de 2003. Ver www.bogor.indo.net.id/kri/bhist.htm.

42. Centro Internacional de Agricultura Tropical (CIAT), "Cassava Boom in Southeast Asia". Boletín *online* del Grupo Consultivo de Investigación Agrícola Internacional (GCIAI), Vol. 4, No. 3, (junio de 1997). Ver www.worldbank.org/html/cgiar/newsletter/june97/9ciat.html.

43. Ver *Processing of Cassava*, Agriculture Services Bulletin No. 8, United Nations Food and Agriculture Organization (Rome: FAO, 1971). Ver www.fao.org/inpho/vlibrary/x0032e/X0032E01.html.

44. Klanaron Sriroth, Kuakoon Piyachomkwan, Kunruedee Sangseethong, and Christopher Oates, "Modification of Cassava Starch". Informe presentado en la décima convención internacional sobre el almidón, Cracow, Polonia, del 11 al 14 de junio de 2002. Ver www.cassava.org/Poland/Modification.pdf.

45. Ver "Improved Germplasm", sitio Web del Centro Internacional de Agricultura Tropical (CIAT), Bogotá, Colombia. Ver www.ciat.cgiar.org/improved_germplasm/cassava.htm, 2001.

46. "Championing the Cause of Cassava", *News & Highlights*, publicación *online* de la Organización de las Naciones Unidas para la Alimentación y Agricultura (FAO), 26 de abril de 2000, 1. Ver www.fao.org/NEWS/2000/000405-e.html.

47. Reinhardt Howeler, agrónomo del CIAT, citado en "Cassava in Southeast Asia: From Hard Times to Modern Times". Op. cit.

48. "What to do About Whitefly?" Artículo publicado *online* por *New Agriculturist on-line: Reporting Agriculture for the 21st Century* (octubre de 2002). Ver www.agri.co.new-UK/02-3/newsbr/html. Ver además Mike Crawley, "Restoring Cassava Production in Uganda". Artículo publicado online en *REPORTS: Science from the Developing World* (Ottawa, Canadá: Centro Internacional de Investigaciones para el Desarrollo (IDRC por sus siglas en inglés), 12 de noviembre de 1999. Ver www.idrc.ca/reports/read_article_english.cfm?article_num=558.

49. Myambu Banea, Nigel H. Poulter, y Hans Rosling, "Shortcuts in Cassava Processing and Risk of Dietary Cyanide Exposure in Zaire". Artículo publicado *online* en *Food and Nutrition Bulletin* 14:2 (junio de 1992) (Tokio: Universidad de las Naciones Unidas). Ver www.unu.edu/unupress/food/8F142Eoa.html.

50. The Cassava Cyanide Disease Network con base en la Facultad de Botánica y Zoología de la Universidad Nacional de Australia en Canberra, Australia, monitorea el envenenamiento y mantiene conectados a científicos de varios países. "La red ha advertido que un procesamiento inadecuado y la sobredependencia de la yuca ha conducido al envenenamiento generalizado en partes de África.... Los métodos tradicionales de procesamiento practicados en el África Oriental, tales como el secado al sol y fermentación en montones, dejan gran cantidad de veneno en la harina de yuca, en contraposición al método de tostado utilizado en África Occidental, que elimina la mayor parte del cianuro. También ha recomendado apoyar una mayor diversidad de cultivos básicos en áreas dependientes de la yuca". En "Fresh Evidence in Cassava Poisoning". Artículo publicado *online* por *New Agriculturist on-line: Reporting Agriculture for the 21st Century*.

51. Los precios promedios anuales a nivel mundial de los gránulos de yuca comenzaron un descenso sostenido en 1996 y continuaron en baja durante todo 2000 debido a la fijación de precios competitivos de granos y a una menor demanda por parte de la Unión Europea. De: "Cassava in Southeast Asia: From Hard Times to Modern Times". Op. cit.

52. "Championing the Cause of Cassava". Op. cit.